The Silent Intruder
Surviving the Radiation Age

The Silent Intruder
Surviving the Radiation Age

Charles Panati
and Michael Hudson

Houghton Mifflin Company Boston 1981

Library of Congress Cataloging in Publication Data

Panati, Charles.
  The silent intruder.

   1. Radiation — Safety measures.   2. Radiation — Physiological
effect.   3. Radiation — Toxicology.   I. Hudson,
Michael, joint author.   II. Title.
RA569.P27   363.1'79   80-25398
ISBN 0-395-29478-9

Printed in the United States of America

S 10 9 8 7 6 5 4 3 2 1

To our mother and father

Contents

Introduction   ix

I X Rays: How Safe?
The Risk to Ourselves   3
The Risk to Our Children   29
Reducing the Risks   43

II Nature's Own Radiation
Our Natural Background Exposure   57
Protecting Ourselves from the Sun's Ultraviolet Rays   68

III The Blanket of Electrical Smog
Microwaves, Radiowaves, and Radar   87
ELF Waves, High Power Lines, and the "Weekend Effect"   108

IV Radioactive Garbage
The Deadly Pileup: Where Can We Put It?   117
Radiation from the Sky: Bombs and Fallout   144

V Nuclear Power
Should It Have a Place in Our Future?   167

Appendices   201

Introduction

There is no denying that radiation has improved and simplified our lives in numerous ways. Lifesaving x rays and medical techniques, modern conveniences like microwave ovens, CB radios, television, telecommunications satellites are all possible because of radiation. However, radiation can also maim and kill, causing cancer in us and birth defects in our children. While enjoying the benefits of radiation, we have ignored these dangers. Almost unnoticed, radiation has become a pervasive, life-threatening pollutant — the invisible killer we cannot see, feel, or smell. Along with poisonous chemicals, radiation has emerged as a major health hazard, and if unchecked, it will become *the* primary pollutant in our lifetime.

X rays are a prime example of the extent of the problem. This year more than 140 million of us will be x-rayed — nearly 70 percent of the population. This is an alarming fact considering that medical authorities admit that 50 percent of these x rays are unnecessary, and that such x-ray abuse is increasing the incidence of cancer and birth defects.

At the same time, we are being bombarded by ever greater amounts of radiation from electrical devices. Today, in this country alone, there are some thirty-six million industrial electrical devices, thirty million CB radios, nine million broadcasting transmitters, ten million microwave ovens, 300,000 microwave broadcasting towers, 50,000 radar dishes, 60,000 miles of overhead high power lines, plus thousands of other radiation-emitting devices such as walkie-talkies, garage door openers, burglar alarms, emergency highway call boxes. Some of the radiation emitted by these devices is known to be harmful. But there is new and disturbing evidence that certain kinds of this radiation may also be capable of causing the same bodily harm as x rays, as well as influencing our behavior — altering our perception and interrupting our sleep.

If this electromagnetic radiation were visible to the naked eye, we would see the eerie glow surrounding department store exits that is produced by electronic frisking devices used for detecting shoplifters. And we could watch shafts of "light" relaying telegrams from the roof of one skyscraper to another, or the beacons of radiation transmitted from airports that guide planes to safe landings. The amount of radiation from such sources has mushroomed to the point where it has actually changed the way the earth looks from space. If in 1900 you could have viewed the earth from space through a telescope sensitive to these radiations, you would have seen a black sphere. Today the earth glows like a white sun, because of the blanket of radiation smog.

In addition to these sources of radiation, the nation's nuclear reactors are churning out millions of tons of deadly radioactive waste. On numerous occasions this debris has seeped into our rivers and streams, with possible grave consequences for all surrounding life forms. Incredibly, after thirty-five years into the nuclear age, our government has yet to develop a safe means of disposing of this radioactive garbage. As well, there is potential radiation danger to hundreds of thousands of us that might result from a nuclear power plant accident.

These are only some of the more dangerous and possibly lethal sources of radiation in our environment.

At first glance, it appears there is nowhere to run or hide from this rain of radiation. Yet we are not defenseless. There are measures for achieving a delicate balance between the wise use and reckless abuse of radiation. These include personal steps that each of us can apply to our daily lives, whether in regard to medical or dental x rays, microwave ovens or other devices, as well as national efforts to tighten the country's radiation standards and reduce the sources of radiation and our exposure. If we do not move promptly and prudently, there is no doubt that the carcinogenic time bomb we are setting will produce a fallout to dwarf the disease and destruction, the mental and physical scars that have become the legacy of chemical pollution.

I X Rays: How Safe?

The Risk to Ourselves

Why the Concern over X Rays?

X rays are a form of ionizing radiation, and ionizing radiation damages our cells by ripping electrons off atoms, endowing them with erratic behavior that can lead to cancer or genetic defects passed on to our children. In general, x rays are produced by manmade electrical devices such as the typical machines used by doctors. X rays also exist naturally. Stars, including our own sun, emit x rays, and extremely hot stars and other celestial bodies can be potent sources of x rays. But these natural x-ray sources do not concern us, since they lie at remote distances in space, and their emissions have no apparent effect on us. It's manmade x rays that have become such an important and controversial part of our modern life.

Few of us would argue that x rays have not provided us with better health care. Yet, because of their indiscriminate use, x rays have become a health hazard as well. Over 140 million of us receive diagnostic x rays for medical and dental purposes each year. That's almost two-thirds of the population. It's no wonder that x-ray radiation has become the largest single source of manmade radiation, accounting for a full 45 percent of all we get.

And we're paying a price. By reliable estimates, nearly 14,000 of us are disabled and another 7500 of us die every year as a result of cancer and other diseases caused by x-ray exposure. The figures

are disturbing, especially considering that about half of the diagnostic x rays prescribed are unnecessary. Little did the early pioneers of radiology realize that, less than a century after their discovery, x rays — once esteemed as the panacea for practically
every malady — could pose such a serious health threat.

Who Discovered X Rays?

A German physicist named Wilhelm Conrad Roentgen. In 1895,
while Roentgen was studying the behavior of a high voltage tube
filled with gas, he noticed mysterious emissions from the tube that
could blacken photographic film — even when the film was
wrapped in several layers of paper and enclosed in a wooden box.
Roentgen immediately began experimenting with the penetrating
power of these rays, photographing the internal structures of normally opaque objects. In one experiment he persuaded his wife to
place her hand between a high voltage tube and a photographic
plate. When he developed the plate he produced a picture of the
skeletal structure of her hand. Roentgen called these mysterious
emissions x rays because he could not understand what produced
them.

It didn't take long for the medical profession to realize the value
of "Roentgen Rays," and before the end of the century, a mobile
x-ray unit was being used by British military doctors in Sudan to
help locate bullets and shrapnel in soldiers wounded in battle.
Since then, of course, medical use of x rays has increased dramatically. In this country in 1979, for example, more than 86 million
of us received dental x rays, while 161 million of us were x-rayed
for medical diagnostic purposes.

Are Today's X Rays Different from Roentgen's Rays?

Not really. When an x ray is taken, the machine used employs the
same basic principles as Roentgen's prototype. When the radiologist places you under or in front of the x-ray machine here's
basically what happens: inside the machine, a stream of fast-mov-
ing electrons is produced and then collides with a metallic target —
usually tungsten — in a vacuum tube. When the electrons hit the

target they are abruptly stopped, causing them to release their energy in the form of x rays. The x rays pass from the machine through the part of your body being examined and fall on the photographic plate placed under it, which records the picture. By varying the metallic target and the speed of the electrons, radiologists can generate x rays of differing "hardness." In other words, they can make them more or less penetrating. Generally, the shorter the wavelength, the harder and more penetrating the radiation. X rays used to treat deep-seated tumors, for example, are hard, whereas dental x rays are soft. The piercing power of x rays also depends on what part of our body is being irradiated and the elements that compose various body tissues. Our flesh and soft tissue like the lungs and liver are easily penetrated, but denser substances like our bones and teeth are more opaque.

Some of our body's dysfunctions or ailments are easier to photograph than others. A fractured bone, the presence of stones in the kidneys or gall bladder, or a calcified tumor are usually visible in a radiograph, often even to an untrained eye. But problems like ulcers, watery tumors, or obstructions and constrictions in the soft tissues of our digestive tract are harder to spot. That's why we have to swallow an opaque solution like barium, which makes the growth or malfunction visible. For the same reason, radioopaque materials are injected into our blood stream so the penetrating x rays can detect abnormalities in our veins and arteries. It is this ability of x rays to penetrate our flesh and bones that makes their application in dentistry and medicine an invaluable lifesaver. But this same property contributes to the potentially lethal aspect of x rays. Unfortunately, the dangers of x rays went unnoticed for decades — despite the many early warning signs.

What Were the First Hints of Danger?

As early as 1906 many radiation researchers had developed skin burns and ulcerations that refused to heal. Some of them suspected the ionizing radiation from x rays as the cause and that x rays could pose a serious health threat. In addition, biologists had irradiated frogs and tadpoles and observed that x rays could cause hideous mutations, and in large enough doses even kill. We may find this hard to believe today, (but virtually all of the pioneering radiation biologists were convinced that the main danger from x rays was inheritable genetic mutations.) They did not suspect x rays could cause cancer twenty to thirty years after a person had been irradiated. In fact, x rays were considered to be the long-sought cure for cancer, and only two years after their discovery doctors in France, Sweden, and Great Britain began using x rays to combat the disease.

As it happened, x-ray therapy did kill tumors, and it wasn't long before x rays were being used as a panacea for practically all illnesses. By the end of World War I, doctors were administering x-ray therapy to large numbers of people to cure everything from enlarged or inflamed tonsils, adenoids, ringworm, acne, and various skin disorders, to the removal of unwanted body hair. X rays were also administered to alleviate asthma, whooping cough, and a form of deafness caused by enlarged lymphoid tissues around the eustachian tubes. Perhaps most misguided of all, x rays were used to shrivel enlarged thymus glands in youngsters to ensure that they would grow and develop into normal adults. These treatments usually directed at the head and neck, were common through the 1950s. In fact, radiation of the thymus in infants and children was a common practice until the mid 1960s. In all, a total of nearly one million people, including hundreds of thousands of infants and children — who are especially prone to cancer from radiation due to their active cell growth — received these dangerous treatments. Eventually, many of them died of cancer.

Didn't Doctors Realize the Harm They Were Doing?

Unfortunately not, since most radiation injuries take decades to surface. So strong evidence that x rays could cause serious long-term damage began to accumulate slowly only around the middle of this century. One of the first people to document the harmful effects of x rays was Dr. Herman C. March. After studying the obituaries in the *Journal of the American Medical Association* for fifteen years, March reported in 1944 that radiologists who regularly worked with x-ray machines were ten times more likely to die from leukemia than other physicians. It didn't take long for the National Cancer Institute to reach a similar conclusion. Gradually it became hard to ignore, or deny, the facts about x rays. Many of the early pioneers of radiation therapy had already died from cancer of the skin and bones. The great physicist Marie Curie, who was the first to isolate radium and plutonium, which emit x-ray-like gamma rays, died of leukemia as did her daughter Irene, who helped her mother in research. By the early 1950s, articles in medical journals were reporting increased incidences of cancer in those people who had received radiation treatments up to thirty years earlier for thyroid trouble and a host of other ailments.

Unfortunately, this research uncovered only the tip of the iceberg. Each year the indiscriminate use of x rays took a mounting toll. The jump was reflected in a National Cancer Institute survey which showed that thyroid cancer rates in this country increased from a 1947 average of 2.4 percent per 100,000 people to 3.9 percent per 100,000 in the late 1960s.

And scientists began realizing that x rays weren't the only problem. Factory women, for example, who in the 1920s hand painted luminous watch dials with paint containing radium, and often licked their brushes to produce a sharper point, began to die of radiation diseases. Eventually almost every factory worker located died of either anemia or bone cancer. Later, leukemia, which shows up faster than most other radiation-related diseases, began to take its toll among survivors exposed to large blasts of ionizing radiation during the atomic bomb blasts on Nagasaki and Hiroshima. By the 1960s the dangers of radiation had become clear to all. But by that time, what was most disturbing was the fact that even a moderate dose of x rays, thought by many experts to be completely safe, could actually result in disease twenty to thirty years later.

Why Does Radiation Damage Take So Long to Develop?

For one thing, (it can take years for radiation-damaged cells to multiply to the extent that they overtake the healthy cells in our organs or body tissues.) The kind of x-ray radiation that most of us receive for diagnostic medical purposes damages our cells either temporarily or permanently. It does not necessarily kill them. You might suppose that cell damage is preferable to cellular death. But there is reason to believe that cells killed outright by x rays present less of a danger to us than maimed ones, because dead cells cannot reproduce or turn cancerous.

On the other hand, permanently damaged cells can propagate their defects(Medical x-ray exposure can damage a cell's chromosomes, which contain our genes, or the nucleic acid, DNA, within our genes, which is the basis for all hereditary traits. These subtle changes can adversely alter a cell's behavior, and as it divides and subdivides the millions of new cells produced will carry the defect and continue to pass it on as the proliferate. After many years the damaged cells finally accumulate to the point where they endanger our health, usually in the form of cancer.)

No one can be sure just how many of our cells are damaged irreparably by exposure to x rays or to other kinds of ionizing radiation. Such speculation is like a game of Russian roulette. Sometimes the x-ray trigger is pulled and no permanent cell damage results, while other times the radiation bullet hits its target. Of course each time we are exposed to radiation, our odds for losing increase. That's not only because of the law of averages, but also because x rays can have a harmful cumulative effect on us.

Radiation Accumulates in Our Bodies?

No, this is a common fallacy. The burst of radiation we get from an x ray is a brief, one-time affair. The only radiation that permanently remains in body tissues is the kind emitted by biologically long-lived radioactive elements. These are administered to some patients for medical purposes. They are also generated as by-products in the production of atomic weapons and nuclear power and can work their way into our food chains and ultimately into our bodies.

However, when we get an x ray, all the potentially damaging energy entirely dissipates in a fraction of a second. What can, and often does, accumulate in our body from x-ray exposure is cell damage. Repeated bursts of radiation, especially at fairly close intervals, can lead to a build-up of potentially harmful irreparably damaged cells and can also prevent the recovery of cells only temporarily damaged. While many damaged cells repair themselves in seconds (or less), others can require much more time. Thus, frequent exposure to x rays obviously increases our chance for permanent cell damage of the kind that can cause cancer.

Do All Irreparably Damaged Cells Cause Cancer or Disease?

Not at all. For instance, irreparable radiation damage to our nerve or muscle cells is less likely to cause us a problem later in life than irreparable damage to our blood cells. The reason is that once we have physically matured, nerve and muscle cells reproduce only infrequently, so the damage can't be transmitted quickly to larger colonies of cells. Blood cells, on the other hand, divide rapidly and can easily propagate a harmful genetic defect, like leukemia. But even an irreparably damaged fast dividing cell may not necessarily cause us harm. In general, what most likely determines harmful cellular destruction is the specific molecular bonds that are altered or ruptured and the particular damage done to the atoms that compose the cell's nuclei. More specifically, damage at a cellular level is a very individual response, especially when the radiation doses are relatively small or modest. There are important personal factors like our age, health, sex, and even ethnic background that play a part in our susceptibility to radiation damage. The age factor, for example, is the reason why children should avoid yearly dental x rays.

Aren't Dental X Rays Harmless? — children

Dental x rays are considered soft, and used prudently they are relatively safe. Yet as with all x rays there is a potential health risk, especially for children who are particularly prone to all kinds of radiation damage. For one thing, children depend on normal cell division to support their rapidly growing bodies, and radiation can impair cell reproduction. In addition, because a child's cells are always busily proliferating, any severely damaged cells will manifest themselves sooner. Also, children have a longer remaining life span than adults, so the ill effects of radiation, like leukemia, have more time to develop and affect them. And since children are small, it is often more difficult to protect their reproductive organs from the primary radiation beam. This is important, since damage to the sex or germ cells can result in hereditary genetic defects.

For all these reasons, very young children who still have their baby teeth should never receive dental x rays on a yearly basis as part of a routine checkup. Their teeth will naturally be replaced in a few years, so it is pointless to directly expose their gums, head and neck to unnecessary amounts of radiation — and possibly expose even their gonads to scattering x rays. Children with their permanent teeth, adolescents and adults, particularly those with a history of cavities, can derive benefits from occasional whole mouth x rays. But regardless of our age, yearly dental x rays are generally unwise.

That Sounds Overly Cautious.

It isn't. In fact, according to the American Dental Association, x rays should not be a standard part of every dental examination. Prudent dentists, the association claims, should require full mouth x rays every five years or so, and take single x-ray films in between — and then only when there are clear indications of a problem.

The caution is understandable. Every time your dentist takes an x ray of your gums or teeth, the skin closest to the x-ray machine receives about 1000 millirems of radiation — ten times the amount of annual radiation our body receives from natural sources like cosmic rays from space and radioactive elements in our air and soil. A millirem is simply a measure of the amount of radiation absorbed by tissue or bone. In a full mouth examination, you may require eight separate films — sometimes even more — which exposes your entire oral cavity to a radiation dose of 8000 millirems. Many dentists perform this procedure on patients twice a year, thereby exposing the oral cavity to 16,000 millirems — more than twice the whole body dose received by the residents of Utah from fallout during the 1950s A-bomb tests. Of course, the dental x-ray dose is not as dangerous because it's delivered only to a small part of the body.

Still, 16,000 millirems of radiation a year is an unnecessary risk to take just to be reassured semiannually that your teeth are in good condition. This is particularly true when you consider that in adults 13 percent of the active bone marrow, which produces our red blood cells, lies in our heads. Bone marrow is one of our body cells most easily damaged by x rays. So even if you are cavity prone, yearly or semiannual sets of whole mouth x rays are generally unwise. It's also important to realize that at times your exposure to dental x rays can even be much higher, since many machines vary in the amount of radiation they give off.

Is This Because Some X-Ray Machines Function Improperly?

Yes, for one thing, many dental machines fail to meet government safety standards. There are about 150,000 dental x-ray units in operation in this country, and according to a recent report from the Bureau of Radiological Health, 36 percent of these machines fail to meet government standards for safety exposure. In fact, dental x-ray machines vary more than any other medical device in the amount of radiation they emit. There are several reasons. Often the units are old models so they don't contain many of the new safety features characteristic of most hospital x-ray devices. In addition, many machines are not calibrated regularly and therefore can often emit more radiation than the control dial indicates. Also, most dental x-ray machines are privately owned, so there is little pressure on a dentist, or incentive for him, to have his machine periodically checked for defects or to purchase a costly improved model.

A further problem arises from the fact that many dentists, especially those trained before the 1960s, continue to regard dental x rays as harmless, because they constitute lower levels of radiation than many other x-ray procedures. As a result, they ignore such safety measures as draping a protective lead-rubber apron around a patient's neck and lap during x-ray exposure to protect the thyroid gland, chest cavity and sex glands. According to a recent

American Dental Association survey, only one in four dentists uses lead-rubber aprons.

The importance of protection for the sex glands in all types of x-ray procedures can't be exaggerated. While the majority of radiation to the sex glands during dental x rays is caused by radiation scatter, the radiation can be more direct. For instance, if your dentist aims the x-ray beam down through your top front teeth, the beam may directly strike your reproductive organs. True, with soft, low penetrating dental x rays the overall danger to the body is slight, yet radiation of the germ cells in our sex glands can result in birth defects so you should insist on adequate protection during dental x rays.

If My Dentist Orders X Rays, How Do I Know I Really Need Them?

First, find out why your dentist wants an x ray. If it's to examine an ulcerated tooth, a diseased root, or some other suspicious symptom that could cause you to lose healthy teeth or result in illness, the x rays are certainly worth having. But if it's merely for the purpose of a routine checkup — which most are — here's a list that can help you make the decision, and some measures you can insist on to minimize your exposure.

- Ask your dentist whether films are absolutely necessary or whether he is fairly confident from his visual examination, and your own personal report, that your teeth are healthy.

- If your dentist insists on full mouth exposures yearly or semiannually, and many of them do, you would be better off changing dentists. Even if you are cavity prone, a full mouth set of x rays once every two years should be sufficient, and some radiologists feel even this may be too frequent.

- Never permit a dentist to use an x-ray machine which is not fitted with a cone. The cone confines the x-ray beam to the local area being x-rayed. Without a cone, your entire head, along with much of your body, is doused by x rays. In addition, if your dentist still uses the old-type plastic pointed cone, ask him why he doesn't have the newer lead-lined cylinder type. It's not very expensive to install, and it greatly reduces radiation scatter.

- Insist that a lead-rubber apron be draped over your body from your neck to beyond your gonads when you're being x-rayed. This will shield you from the unavoidable radiation that scatters from even the most efficient machines.

- Remember that fetuses are particularly vulnerable to radiation damage. So a pregnant woman should avoid dental x rays altogether — even using a lead-rubber apron since, although the chance is slight, the apron could slip off just at the moment her teeth are being x-rayed. And if a husband and wife are planning to have a child, the woman would be wise to visit her dentist before conceiving in case she requires dental x rays.

- If your dentist will not listen to your questions or requests, do yourself a favor and find another dentist.

- If you change dentists, or are referred to a specialist, ask your previous dentist to forward your x-ray records. This information aids the new dentist in understanding your dental history without adding to your x-ray exposure.

Your risk of developing cancer from dental x rays that are prudently taken is very slight. Statistically, for every one million people who receive a whole mouth x-ray series, one person can be expected to die from either thyroid cancer or, more likely, from leukemia. Expressed another way, you run the same risk of developing cancer from a single whole mouth dental x-ray exam as does a person who smokes one cigarette a day for one year. Naturally, if you smoke and have x rays your risk is higher.

Suppose My Doctor Insists on a Chest X Ray Every Year?

Don't take his advice. A chest x ray can save your life if it uncovers a benign or malignant tumor, penumonia, or other chest or lung disease. So if you experience any persistent disturbing or abnormal symptoms in your chest, an x ray is not only wise but probably necessary. But a chest X ray every year as part of a routine physical checkup is foolish and risky. Even if you are a heavy smoker, and therefore in a higher cancer risk category, yearly x rays are too frequent. It would be wiser to schedule them two to three years apart.

In terms of dose and risk, chest x rays fall into the same general category as x rays for the whole mouth, neck, upper spine, shoulder, skull, and extremities like our arms and lower legs. These are all considered to be low dose in nature. Still, according to organizations like the American College of Radiology and the American Medical and Dental Associations, none of these x rays should be taken unless there are clear-cut reasons to believe they will contribute to the diagnosis of an ailment.

There is reason for caution when you consider that every time you get a chest x-ray examination, which usually requires two films, you receive about 90 millirems of radiation to your skin, 15 millirems to your blood-producing organs, and somewhat less to your gonads. Statistically speaking, if you receive just one chest x-ray examination in your lifetime, the chances of dying from cancer are one in one million if you're a man, and perhaps twice that if you're a woman, because of the additional risk of developing breast cancer. Granted, these represent extremely small risks, but as with all diagnostic x rays, the risks are only worth taking when they are clearly outweighed by positive health gains.

Under What Conditions Is a Chest X Ray Justified?

Here are several important considerations in determining whether or not you should have chest x rays and, in some instances, low-dose x rays in general.

• If your doctor insists on yearly or frequent chest x rays as part of a routine physical, he is not practicing good medicine, and you would be better off choosing another doctor. This is especially true if you generally enjoy good health, or if you are a young adult, and as such have a lower risk of lung disease than older people.

• If your doctor orders a chest x ray, ask him first to consult previous recent chest x rays if you've had them. Physicians may fail to use current films because they just can't be bothered sending for them, or because they are unaware — especially if you were previously under the care of another doctor — of their existence. The results of previous x-ray films can spare you unnecessary radiation.

- If you are expecting to undergo major surgery, the hospital will require a chest x ray as part of a series of tests to ensure you are in good health. However, if you already had a chest x ray in the past year, showing you are in good health, and you are undergoing routine surgery, such as hernia repair or appendectomy, 99 percent of the time there is no need for another x ray. Often you can avoid the unnecessary radiation by mentioning your recent x-ray exam to the surgeon.

- If your doctor orders an x ray and you are in doubt, obtain a second opinion. However, because of the casual regard with which many doctors treat x rays, you may well get the same advice.

- Whatever you do, don't request an x ray yourself. Often when a patient expresses concern about a discomfort and seeks assurance that it is not cancer or another serious disease, a doctor will x ray merely to allay the patient's fears. Sometimes doctors will x ray when they are confident there is no disease because they fear a malpractice suit might arise if the patient develops an ailment and claims the doctor neglected to treat the symptoms.

Are Fluoroscopes Safer than Chest X Rays?

Definitely not. In fact, your exposure to radiation is even greater — from two-and-a-half to fifty times more than from the standard chest x ray. Fluoroscopic procedures are in a relatively high x-ray radiation dose category, along with x rays of the upper and lower gastrointestinal tract, gallbladder, kidneys, ureter, bladder, breasts, lower and midspine, pelvis, hip, and upper thigh. Obviously it is even more important for you to question a doctor about the necessity of any of these procedures.

As for fluoroscopy, it is based on the same principles as radiography, but instead of producing a permanent image on film, the process yields an immediate, temporary image on a fluorescent screen, which glows like a television screen when it is struck by radiation. If you've ever had a fluoroscope, you know that during the procedure you stand between the x-ray machine or other radiation source and the fluorescent screen. The radiation passes through your body, producing images on the screen which are composed of varying degrees of light and shadow. To produce this on-the-spot movie you have to be exposed to radiation for a relatively long time.

Then Why Is Fluoroscopy Used?

Because the procedure has two advantages over radiography. It is less expensive and requires less time, since the doctor immediately and continuously views the part of your body under examination, without having to wait for a photographic plate to be developed. But this very immediacy of a fluoroscope, and the fact that it doesn't employ a cone to direct the large amounts of radiation at a target area, are precisely what make exposure from a fluoroscope more dangerous than from a standard x-ray machine.

Despite the increased risk, there are certain medical situations which require a physician to see a live image, or size, shape, and movement of a patient's internal organs. The decision to fluoroscope or to use a series of x-ray films should be made by a competent radiologist. Unfortunately, because fluoroscopes can generate instant on-the-spot examinations, there is often a tendency to resort to them when they are not needed. If your doctor exposes you casually from time to time to a fluoroscope to check for congested lungs or other ailments in the chest cavity, he is performing a great disservice.

Is There Any Proof Fluoroscopy Can Be Harmful?

If you question the potential dangers of fluoroscopes, consider these facts:

- In the 1940s, in a sanatorium, a group of 300 women suffering from tuberculosis received several routine fluoroscope examinations as supposedly harmless checks on their disease. Twenty-two women later developed breast cancer, compared with only two cases of the disease in the women not fluoroscoped. Women in two other TB sanatoria between 1930 and 1956 were periodically examined with fluoroscopes. Subsequently, they developed 80 percent more breast cancer than women who did not receive such examinations. The doses in all these instances were high. Nonetheless, they constitute proof of the potential danger, particularly breast cancer, from fluoroscopes.

- On a level more relevant to all of us, in the 1950s the government sponsored a massive preventive disease program aimed at detecting the early stages of lung and heart disease. Under the program millions of Americans were x-rayed and fluoroscoped in local mobile medical units. More than a decade later, the program was abruptly halted when scientists began to realize that the dangers posed by x-ray screening programs for the general public invariably outweigh the benefits. Unfortunately, though, health officials didn't learn enough from their past mistakes.

Why?

In 1973 the National Cancer Institute (NCI) and the American Cancer Society (ACS) sponsored a National Breast Cancer Detection program which employed mammography, or breast x rays, as one of its screening techniques. As it happens, the breast is extremely vulnerable to radiation. The program was hailed as an exercise in preventive medicine aimed at detecting breast cancer in its early treatable stages. Twenty-seven medical centers participated in the program along with several hundred thousand women. In addition, in 1974, millions of women across the country sought mammographies to ensure both their physical and psychological well-being following the much publicized mastectomies of Betty Ford and Happy Rockefeller.

Many of the mammographies under the NCI-ACS program

were performed by doctors not trained in radiology who used uncalibrated or obsolete equipment. Ralph Nader's Health Research Group discovered that seventeen x-ray machines being used in the NCI-ACS program were delivering unnecessarily high levels of radiation. One machine at Georgetown University was giving off radiation three times the already high prescribed maximum dose of about 1500 millirems. Worst of all, many of the women who submitted to mammographies were in their twenties, and women in this age group are not good mammography candidates for two reasons. First, the incidence of breast cancer is much lower in women under age thirty-five. Second, for hormonal reasons, it's much harder to obtain images of breast cancer in premenopausal women; breast films of young women in the hands of all but the most experienced radiologists are practically worthless.

The widespread use of mammographies under such conditions was nothing less than scandalous and a clear example of the abuse of x rays. A large number of perfectly healthy women received appreciable doses of unnecessary and harmful radiation. This is particularly frightening since the risk for radiation-induced breast cancer is higher than that for any other malignancy: if one million women each receive 1000 millirems of x rays, between 50 to 200 can be expected to develop breast cancer. (If you're interested in comparing this with the cancer risks to other organs from x rays, consult Appendix B on page 204.) The latency period, or time between exposure and the appearance of the disease, is from twenty to thirty years, though it can be as short as ten years.

So Some Women Who Participated in the Screening Will Almost Surely Get Cancer?

Yes, probably in the 1980s or 1990s. In fact, in 1975 the National Cancer Institute examined all the available evidence on the effects of mammography and concluded that repeated mammograms in women under age thirty-five, who are not high cancer risks, are extremely dangerous and probably cause more cancer than they detect. They also recommended that the routine use of mammography in screening women between the ages of thirty-five and forty-nine who had no symptoms of the disease be discontinued.

As a result, the NCI has ended routine screening by x-ray mammography for women under fifty, except in cases where the benefits are believed to outweigh the risks, such as women with histories of breast cancer either in themselves or their mothers or sisters.

For women over age fifty regular mammography every several years may be wise because postmenopausal women run a higher risk of breast cancer. Fortunately, abnormal growths in the breasts are more easily detected in this age group. Women over fifty run less of a risk from breast x rays for another reason as well. The few who may experience serious cell damage as a result of breast x rays will probably die of natural causes before the damage manifests itself in twenty to thirty years. This age factor is an important consideration for women, as well as men, when weighing the risks of all kinds of x rays.

Routine Breast X Rays May Be Safer for Older Women, but Haven't They Saved the Lives of Many Younger Women?

They certainly have, and considering that breast cancer is the number one killer of American women, mammography is a crucial asset in the health care of women. But a series of films in a routine mammogram can expose healthy breast tissue to an average of 4500 millirems, and, in some cases involving more extensive examinations, to 10,000 millirems. So the procedure should not be casually administered on a wholesale basis. Statistics show that, for every one million women who receive one mammography examination, five to twenty can be expected to die from radiation-induced cancer.

Regardless of age, however, if you develop any of the warning signs of breast cancer, especially if you are hereditarily disposed to it, you should have a mammogram. The possibility of a metastasizing malignancy is certainly more to be feared than the health risk from a breast x ray. In this circumstance, there is no reason to be fearful of the risk. Since the breast is composed entirely of soft tissue with no bony structures, it can be x-rayed satisfactorily with special low-voltage machines. These can illuminate small differences in the composition of muscle and fat and highlight small irregularities that could indicate a benign cyst or a malignant tumor.

An equally effective method for detecting breast growths is xeromammography, which delivers about the same x-ray dose as a mammogram. This process employs a form of mammography in which the breast image is recorded by an electrostatic method, similar to that used in Xerox copying machines, rather than on the more conventional film. Regardless of the method of breast exam you may have to undergo, there are safety precautions available that can greatly minimize the radiation risk.

What Are the Safety Precautions for Mammography?

• If you are over fifty, or a younger woman with a history of breast cancer, or if your mother and sisters have such a history, you should choose a large hospital or university medical center for a mammogram rather than a small private doctor's office. A large center is more likely to have installed the latest safety equipment and have a staff of skilled technicians and doctors trained specially to administer and read breast films. In general, all of us receive significantly less radiation exposure at such institutions because the machines are supervised by a full-time radiologist. For example, mobile units with fluoroscopic equipment usually emit five to ten times more radiation than those at a radiology facility.

• You should not schedule mammograms on a yearly basis. An examination every two or three years is preferable since a longer time interval not only diminishes radiation risks but allows any cell damage more time to mend.

• If you are pregnant, absolutely avoid a breast x ray. Only in the most extreme situations where your life or health is clearly at stake should any x rays be administered during pregnancy.

• If you must have a mammogram make sure your midsection is draped with a lead-rubber apron to protect the rest of your body and, in particular, your sex glands.

It should be of some comfort to all women that the Food and Drug Administration, with support from the National Cancer Institute, has set up a voluntary program aimed at substantially reducing the amount of radiation received during mammography. In the program the newest equipment is being used, physicians and technicians are trained in the most effective exposure techniques, and x-ray machines are being calibrated regularly. More than forty states are currently participating in the program, and so far it seems to be paying off. Pilot studies have shown that radiation doses to the breasts can be reduced by an amazing 46 percent. This is encouraging considering the fact that women in general are especially susceptible to the harmful effects of radiation.

Women Are More Vulnerable than Men?

Absolutely. If one million men were exposed over their lifetime to a dose of 1000 millirems of radiation between 12,000 to 42,000 of them would develop cancer. In a population of one million women receiving the same dose, the estimate ranges from 24,000 to 94,000 cancer cases. So women are more than twice as likely as men to develop cancer from irradiation. To put these figures in some perspective, the incidence of cancer of all origins is roughly 250,000 cases per million people over a lifetime.

While the age factor in radiation susceptibility has been recognized for some time — that children are more vulnerable than adults — this sex difference was only recently discovered. Through the early 1970s doctors believed that leukemia was the most likely of all radiation-induced cancers, and that men and women ran essentially equal risks from exposure to "low" levels of radiation, like the kind we receive from medical diagnostic procedures. Now we know that radiation more readily causes so-called "soft tumors" in the breast and thyroid, as well as in the lungs and digestive tract. Because women are more prone to breast and thyroid cancer, they therefore suffer a greater overall radiation risk.

But the role of the sex factor is not so clear-cut. While women appear to be twice as susceptible to radiation-induced thyroid cancer, men, on the other hand, are more prone to leukemia from radiation exposures. Our ethnic origins, too, seem to play a role in determining our vulnerability to radiation damage. Jewish children, for instance, are believed to run a higher risk of developing radiation-induced thyroid cancer than children of other ethnic groups. Doctors aren't sure why such variations exist, though they certainly involve our different genetic make-ups which manifest themselves in such well-established facts that blacks, for instance, are more prone to sickle cell anemia, Mediterraneans to Cooley's anemia, Portuguese to the neurological disease known as striatonigral degeneration, and Jews to Tay-Sachs disease. In addition to ethnicity, sex, and age, another factor is important in evaluating radiation risks — your general state of health.

Are Healthier People More Resistant to Radiation Diseases?

This is generally the case for people exposed to relatively large doses of radiation and it may even be true for exposure to small doses. There is reason to believe that people prone to viruses, colds, and other illnesses are probably more sensitive to the damaging effects of x rays as well as other forms of radiation. Laboratory experiments have repeatedly demonstrated that animals raised in sterile environments and irradiated exhibit significantly greater resistance to all kinds of radiation damage. This is due, at least in part, to the fact that the sterile environment eliminates the possibility of infection, which might proliferate after the body's immune system has been weakened by radiation exposure.

There are indications that diseases, particularly in early childhood, apparently heighten a person's susceptibility to radiation illnesses. Dr. Irwin Bross of the Roswell Park Memorial Institute in New York has demonstrated that children under age four with allergic diseases, such as asthma or hives, have a 300 to 400 percent increased risk of dying of leukemia compared with children who have no allergies. His studies show that children with allergies, who had been irradiated in the womb, have an even higher leukemia death risk of 5000 percent. On the other hand, children with no allergies, who had been exposed to diagnostic x rays in the womb, reflect a 40 to 50 percent increase in the death risk from leukemia. So there appears to be a synergistic, or multiplying, effect between certain illnesses and exposure to radiation.

There is also an indisputable synergistic effect between smoking and radiation-induced diseases. A large body of evidence clearly demonstrates that men who both smoke and work in uranium mines, where they are exposed to naturally occurring radiation, develop lung cancer at an enormously higher rate than either smokers not exposed to radiation or uranium miners who do not smoke. There is growing evidence, too, though not yet conclusive, that if you drink excessively or exist on a very poor diet your susceptibility to the effects of radiation may be heightened. And what's interesting is that, while these bad habits can hasten the aging process, there are indications that radiation can do likewise.

X Rays Can Accelerate Aging?

All forms of ionizing radiation can accelerate aging, though the effects are slight for all but the largest doses. It is estimated that our life is shortened by two to fifteen days for every dose of 1000 millirems to our entire body. The 1000 millirem figure constitutes a very rough estimate, and there is undoubtedly great variability from person to person, especially for children. We know, for instance, that young animals that receive substantial whole body doses of radiation, and survive, look and behave just like older animals that have not been irradiated.

There are two major theories as to how radiation ages us. One theory maintains that the diseases we experience in early childhood, plus the toxic chemicals in our environment, as well as natural and manmade sources of radiation act in concert to weaken our body's immune system. By fighting infections, the immune system is responsible for the general fitness of our body, and some scientists speculate that the immune system may be directly linked to the aging process. If that's the case, impairment to this system can hasten aging. According to this theory, though, manmade radiation, such as from atomic bomb fallout and x rays, is only a secondary factor in aging. The primary factors are a high-fat, high-cholesterol diet, smoking, obesity, and lack of exercise.

The second theory points more directly to manmade and natural radiation as a cause for aging. It maintains that the speed at which we age depends on the number of radiation-induced chromosome aberrations that cannot heal themselves. The greater the number of aberrations as a result of radiation exposure, the more rapidly we age. This theory has some credence in light of the fact that many of the chromosome aberrations believed to make us age faster have been found in the blood and liver cells several months after laboratory animals and patients have been exposed to radiation. In general, the immediate effect of chromosome aberrations, according to this theory, is to impair our normal cellular functions which hasten our body's natural deterioration, leading ultimately to death.

How Much Radiation Does It Take to Cause Immediate Death?

For humans, a single dose to the whole body of 500 rads, or 500,000 millirads, is considered deadly. A rad is the unit that measures the radiation absorbed by our bones and tissues, from *all kinds of radiation;* it is equal to a rem for doses of x rays and gamma rays. Lethal doses are usually what we could get from an atomic bomb blast or a serious nuclear power plant accident, such as the meltdown of a reactor core. The lethal dose is referred to as the LD/50, which means that 50 percent of any large population receiving this amount of radiation could not possibly survive.

It is interesting to note that lower life forms are more resistant to radiation death. The LD/50 for sheep, swine, dogs, and cats is about 220 rads; for monkeys (and humans) 500 rads; for rabbits, mice, and hamsters, roughly 900 rads; and for fruit flies, 80,000 rads. One of the most radiation-resistant creatures on earth is the cockroach. It has an LD/50 of 100,000 rads. In the event of global destruction from nuclear war, there is a high probability that the planet would be left to cockroaches and radiation-resistant bacteria — a scenario that has been explored many times in science fiction books and movies.

It is important to point out that, by definition, to be lethal the LD/50 must be delivered to the entire body. In radiotherapy for the treatment of cancer, patients routinely receive much higher doses to specific parts of their body containing malignancies. The patient not only lives but suffers only minor side effects. That's not to say that the intense radiation of cancer therapy does not carry long-term higher risks than ordinary diagnostic x-ray exposure. However, many cancer victims are middle-aged or older so it is pointless to worry that a treatment which could save their lives might actually cause a malignancy in twenty to thirty years. For children, however, who are more susceptible to radiation damage, radiation therapy is obviously more risky. That's why surgery and drug treatment, such as chemotherapy, may often be preferable in treating childhood malignancies.

In general though, therapeutic doses are not lethal because the radiation is delivered to an isolated part of the body and kills mainly diseased cells (though unavoidably many healthy cells also) in the narrow path of the radiation beam. On the other hand, if our entire body were to receive a lethal dose, the resulting effects would be hideous and almost always irreversible.

What Are the Immediate Effects of a Lethal Dose?

If you were to receive a lethal dose of 600 to 1000 rads from, say, a nuclear power plant accident or a bomb blast, for instance, within four to five hours you would experience loss of appetite, nausea, vomiting, and bloody diarrhea. These symptoms would persist for several days, and in addition you could possibly lose your hair and your fingernails might turn black. The tissues that generate your blood — the bone marrow, lymph nodes, thymus gland, and the spleen — would be seriously affected. Your white blood cell count may decrease from its normal level of 8000 (cells per cubic millimeter) to 50. Consequently, your body would lose its major defenses against all kinds of bacteria, and toxic and infectious substances. This is medically known as the *hematopoietic syndrome* from radiation, and it alone is enough to almost certainly kill you. Usually, it is the means of death for whole body doses between 400 to 600 rads. If you were unlucky enough to survive, you would almost certainly die within five to ten days as the result of another body breakdown — the erosion of the lining of your gut.

The surface of our small intestines are lined with thousands of villi, fingerlike projections covered with special cells that extract nourishment from food. These cells do not reproduce, and are constantly worn away by food passing through. They must be replaced by new cells from the base of each villus. A whole body dose of radiation between 600 to 1000 rads destroys the replacement process, and the villi are worn down to a point where the body is deprived of nourishment. In addition, holes soon appear in the villi and the harmless, essential bacteria in our bowels — which aid in digesting food — escape into the blood stream, infecting our entire body. This is known as the *gastrointestinal syndrome* from radiation.

If you were unfortunate enough to receive a whole body dose greater than 5000 rads, you would most likely die within twenty-four to forty-eight hours from what is believed to be massive trauma to your central nervous system. The sequence of events, known as the *central nervous system syndrome,* indicate you would quickly become disorientated, lose all coordination, and have difficulty in breathing. You'd develop convulsions and within hours lapse into a coma that would end in death.

Have Any People Actually Received Lethal Doses?

Yes, there are about a dozen incidences in which people have received huge lethal doses of radiation as a result of accidents at nuclear reactors or in uranium processing plants. In 1958, a young man working at a nuclear reactor facility accidentally received a whole body dose of x rays, and the even more damaging neutrons, totalling 4000 rads. He went into shock immediately and was unconscious within minutes. After eight hours doctors could detect none of the infection-fighting blood cells known as lymphocytes in samples of his blood. He never regained consciousness and died within thirty-five hours of the accident.

Another serious accident occurred in 1964 to a thirty-eight-year-old man working in a uranium plant. He received a whopping whole body radiation dose of 8800 rads from both x rays and neutrons. He recalled seeing a brilliant flash of light and was hurled backward against a wall and stunned. But he remained conscious, quickly scrambled to his feet and raced from the building. However, as soon as he got outside he complained of intense abdominal cramps and headache; he began to vomit, and within a few hours his stools were soft and bloody. The next day he felt more comfortable, but strangely restless. On the following day he found it difficult to breathe, his vision blurred badly, and his blood pressure plummeted to a dangerous low. That evening he became completely disoriented, lost muscle coordination, and died — forty-nine hours after the accident had occurred. In instances such as these, people in the general vicinity of the accident — but at a relatively safe distance — have reported smelling the burning of human flesh.

These two men died from the more immediate trauma to the central nervous system. But there is one case on record of a man suffering the more delayed gastrointestinal death from radiation exposure. In 1946 a thirty-two-year-old man at a nuclear facility received a whole body dose of x rays and neutrons of 1500 rads. He vomited several times within the first few hours and his temperature and pulse rate rose. Otherwise, his general health remained relatively good for the next five days. Then he suddenly developed intense abdominal cramps and his intestines became severely blocked with body fluids that had to be mechanically suctioned out. Within twenty-four hours 2½ gallons of fluids were removed, and he died two days later. At autopsy doctors found that the inner walls of his intestines had been burned away by radiation and bacteria had invaded his entire abdominal cavity.

Are the Symptoms Basically the Same for Large but Sublethal Doses?

For doses between 150 to 400 rads the symptoms are painful but much less severe. You would probably survive, though the closer the dose is to the 400 mark the slimmer your chances. A person who receives a dose in this range experiences nausea, vomiting, and malaise within the first few days after irradiation. This is followed by an illusory period of relative well-being that can last up to two weeks. Next comes acute anemia and internal hemorrhaging in the body's mucous membranes. If death occurs, it will usually take several weeks and result primarily from infection;

during this time the victim experiences progressive loss of hair and body weight.

For a whole body dose of less than 150 rads you would experience these symptoms in a very mild form. And although you would likely survive, you would run a high risk of malignancies later in life. Also there is a significant chance that you would pass on to your children a propensity for cancer and genetic weaknesses for many diseases. Worse yet, because of radiation damage to your sex glands, your children could be born with gross physical deformities.

The Risk to Our Children

How Do X Rays Cause Birth Defects?

X rays contribute to birth defects in one of two ways: either indirectly, by damaging our sex cells so we produce deformed offspring, or directly, by harming the developing child in the womb. Irradiation to the developing fetus presents the greatest potential for damage. We know this primarily from animal studies and by observing children born to women who received x rays to the pelvic area while they were pregnant.

The most significant animal data come from the "Megamouse Project" at the Oak Ridge National Laboratory in Tennessee involving seven million mice. The study was begun several decades ago to determine if there was a direct link between x rays and birth defects. The unequivocal conclusion is that there is such a link. Over the years mice exposed to x rays have consistently produced offspring with bent or gnarled tails, deformed ears, splotchy fur coloration, and deformed limbs and brains. Like any animal experiments, the results cannot be easily extrapolated to humans. However, the developmental stages of a mouse and human fetus bear many close biological resemblances, and also, the chemical nature of the hereditary material DNA is basically universal among all animals, including man. So there is scientific reason to believe that the mouse data provide a fairly reliable indication of the genetic threat that x rays, and ionizing radiation in general, present to us and our children. In fact, the results of the Megamouse Project have become the basis for most of our safety recommendations for protecting men and women of childbearing age from exposure to radiation, as well as pregnant women.

Why Rely on Animal Studies? Isn't There Any Convincing Human Evidence?

For obvious reasons, there have been no experiments conducted to determine the danger of radiation to the human fetus. However, that does not mean there is no human evidence. For one thing, records of women who have given birth to physically abnormal children show a definite cause and effect between x-ray radiation and deformities. (Several studies have established that children born to mothers who receive radiation to the pelvic region during pregnancy have a much higher incidence of abnormalities, particularly deformed limbs and brain damage.)

(In one study of 106 women who received large therapeutic doses of radiation to combat cancer and other illnesses, thirty-one women experienced abortions. Of the seventy-five women who eventually gave birth, thirty-eight of their infants displayed a wide range of severe deformities: sixteen infants suffered from tiny, underdeveloped brains, two were mentally retarded, one had a swollen brain, and another possessed the features of a mongoloid. Eight children suffered various body deformities like clubfeet and stunted arms, or were blind.) It must be emphasized that these women received large therapeutic doses of radiation, far more than what we get from most diagnostic x-ray procedures. In addition to congenital physical abnormalities, there is also evidence that children irradiated in the womb run a greater risk of developing malignancies either in their childhood or later in life.

How Much Radiation Does It Take to Harm a Fetus or Baby?

While exact numbers are lacking, Dr. Alice Stewart, a British epidemiologist at the University of Birmingham, has found that if a (woman receives only one diagnostic x ray to the abdomen, she increases her child's risk of leukemia by 40 percent.) Other studies are in basic agreement with Stewart and suggest that children irradiated in the womb as a result of diagnostic breast, chest, and upper and lower gastrointestinal x rays run a 50 percent greater risk of dying from leukemia and other malignancies by the time they are sixteen years old.

Despite the relatively low diagnostic x-ray doses that produce these awful effects, as recently as 1975 nearly 7 percent of all pregnant women in this country received pelvimetry x rays to determine the size of their pelvises. The radiation dose to the fetus in the womb from a pelvimetry examination ranges from 600 to 4000 millirems — enough radiation to result in five to twenty spontaneous abortions or stillbirths for every one million pregnant women irradiated. If you're an expectant mother and interested in learning of the amount of radiation to the fetus from various diagnostic x-ray tests, you might be required to have for medical purposes, you can consult Appendix C on pages 204–205.

Of course, x rays aren't the only form of radiation that can cause birth defects. All forms of ionizing radiation can. Studies of the descendants of the survivors of the atomic bomb blasts on Hiroshima and Nagasaki have revealed abnormally high incidences of congenital leukemia. In addition, many Japanese women aborted their babies shortly after the blasts, while others gave birth to children with deformed heads, underdeveloped brains, and grossly deformed limbs.

In short, there is no question about the link between radiation and birth defects, and according to a 1979 report by the International Commission on Radiological Protection, the threat from x rays and all forms of ionizing radiation is much greater than it was once considered to be.

How Much Greater Is the Threat?

For years scientists believed that radiation exposure presented a greater danger to our sex cells than to the rest of our body cells, known as somatic cells. However, they now judge that the ratio of somatic to germ, or sex, cell damage may be sixty times greater than it was once thought to be. This is bad news for us and the developing fetus. In human terms, it means that manmade sources of radiation are responsible for 6000 of the 350,000 deformed babies born in this country each year. These figures, determined by the 1979 government commission on the Biological Effects of Ionizing Radiation (BEIR), are rough estimates and a minority of scientists feel they are five times too high. Still, the numbers are disturbing.

Radiation-induced birth defects are indistinguishable from those that occur naturally, so it's nearly impossible to identify them among the 197,000 serious genetic disorders that occur in every million births from natural and all environmental causes. This doesn't diminish the fact, though, that x rays present a very real danger to the unborn child. That's why only in the most extreme situations, where a woman's life or health is clearly in jeopardy, should x rays be administered during pregnancy. Even if a pregnant woman breaks a bone and x rays are required, extreme precautions should be taken to shield her abdomen from scattering radiation. Such caution is warranted especially in the early weeks of pregnancy because a developing embryo is particularly sensitive to the effects of small amounts of radiation.

Is a Fetus More Likely to Be Damaged at Any One Stage of Development?

Yes it is, and also the radiation it receives at various stages can yield very different abnormalities. From the moment of conception until the ninth day of pregnancy the embryo is not yet embedded in the wall of the uterus. If a woman were irradiated during this time, there is a chance the embryo would be damaged in such a way that it would be aborted.

From the ninth day through the sixth week of pregnancy the fetus begins to develop body organs and limbs in a strict sequence that is universal among humans and many animals. At any one stage of development the cells in some parts of the fetus divide rapidly, while in other parts they divide more slowly. Radiation affects the rapidly dividing tissues more easily. The result is that the pattern of development can be distorted, and the infant will have disproportionate parts. This period of organ and limb development is called organogenesis, and it is at this stage that radiation can produce the greatest deformities.

Radiation biologists, basing their conclusions on animal studies, believe that deformities of an infant's lower jaw, and the occurrence of abdominal hernias, usually occur from radiation received during the 12th to 20th day of pregnancy; ear, nose, and eye deformities from the 20th to 30th day; structural brain abnormalities from the 15th to 25th day; and an infant can develop a cleft palate from being irradiated from the 28th to 34th day of gestation. If you're interested in learning when other deformities are most readily induced by irradiation of the fetus, consult Appendix D on page 206. Unfortunately, it was during this period of organogenesis that the drug thalidomide was administered to many pregnant women in Europe, who gave birth to babies with stunted, and often flipperlike, arms and legs.

Is the Fetus Any Safer After the Sixth Week?

By no means. After the sixth week of pregnancy, when most of the organs and limbs have assumed their basic shape, another danger arises: the sizable risk of producing cell damage that could result in a perfectly formed baby, but one born with leukemia, some other form of cancer, or a propensity to develop these diseases later in life. A fetus irradiated after the sixth week of pregnancy also runs the risks of stunted growth as an adult and a diminished intelligence. There is good reason to believe that in general a fetus is five to ten times more sensitive than an adult to radiation-induced malignancies.

Obviously, once a woman realizes she is pregnant, she can avoid x-ray exposure, unless a test is mandatory because of grave consequences to her health or life. So the real danger to the fetus occurs in the very early days of pregnancy, before a woman knows that she is pregnant and for some reason undergoes a pelvic or stomach x-ray exam.

If a Woman Is Trying to Conceive, How Does She Protect Herself Against X Rays?

There is a measure that can be taken to protect a fertilized egg even at its earliest stage of development. It's the "Ten-Day Rule." According to the rule, if for any medical purpose a woman needs x rays of her stomach or pelvic area — or, to be absolutely safe, any x rays at all — she should schedule them *only during the first ten days of her menstrual cycle,* because naturally there is no fertilized egg in her womb. This sounds like a simple enough safety precaution, but too many times doctors and x-ray technicians ignore it either because of organizational or scheduling problems in hospitals, or simply because it is an inconvenience to them. Therefore, the burden to adhere to this rule lies with the woman.)

Any woman in her child-producing years requiring an x ray of the pelvic area, in particular, should inform her doctor about her menstrual period. If a woman has been scheduled for a pelvic or stomach x ray during the critical twenty-day period, when a fertilized egg may be in the womb, she should reschedule the test until after she is certain she's not pregnant.

Of course, there are bound to be times when the ten-day rule cannot apply. If a woman suffers serious injury in an automobile accident, or if her doctor suspects she has cancer of the stomach, x-ray examinations of the pelvic area may be a matter of life or death. At a time like this, even a woman who is known to be pregnant may have to take the risk.

Every woman should be aware of one additional fact. If she decides to defer what her doctor considers a necessary x-ray examination until after her next menstrual period, the delay may not be merely two or three weeks; if she is pregnant, the delay could be nine months. Most radiologists suggest that there is little sense in deferring necessary x rays during the twenty critical days of the menstrual cycle unless the doctor decides the delay could, if necessary, be extended for nine months without jeopardizing health or life.

If a Pregnant Woman Must Be X-rayed, Is There Any Way to Know if the Embryo or Fetus Has Been Harmed?

There are certain guidelines that can help a doctor determine what kind of damage might have been caused. First, the woman's gynecologist should ask the radiologist for a fairly reliable estimate of how much radiation the fetus received. Knowing the woman's weight, measurements, and the x-ray dose, radiologists have standard means for estimating the dose to the fetus. Second, the gynecologist should attempt to pinpoint as nearly as possible the stage of development of the fetus at the time the x ray was administered. Knowledge of this and the radiation dose to the fetus can aid a doctor in judging the extent of possible damage.

Based on patients' records, many radiologists believe that if the embryo receives a dose in excess of 10,000 millirems of radiation, particularly during the first five weeks following conception, there is a very strong possibility that the child will suffer from abnormalities. While most diagnostic x-ray procedures do not deliver such large doses to the fetus, several other common tests may, depending on the techniques employed by the radiologist. For instance, a barium enema of the large intestines, a KUB (of the kidney, ureter, and bladder), an aortagram (of the heart's aorta), or a celiac angiogram (of the abdominal blood vessels), performed using a fluoroscope, can deliver doses in this very critical range. In addition to these diagnostic tests, there are many forms of radiation therapy for cancer that deliver far more than the critical 10,000 millirems to the fetus. If a woman who is pregnant, or suspects she's pregnant, is exposed to such large doses of radiation, she may very well wish to consult her doctor about an abortion.

However, the 10,000 millirem figure is not rigid, and how much weight it carries in helping shape a woman's decision may vary with the circumstances. For instance, if a woman wants children and for medical reasons another pregnancy is not possible, she may choose to risk having the baby. On the other hand, if the expectant mother is young and fertile, she may find little reason to risk having a deformed child and elect to have an abortion — even if the dose is less than 10,000 millirems. Whatever the case, the risks should be carefully spelled out to the mother and she must be involved in the final decision.

What About the Danger that Deformed Children Will Result from Radiation Damage to the Sex Glands?

There is a genuine lack of strong evidence to support a link between irradiated sex cells and subsequent birth defects, so scien-

tists regard such damage as less of a risk than direct irradiation of
the fetus. This doesn't mean, however, that the risk isn't very real.
(Damage to our sex cells is the only way radiation-induced defects
can be inherited by our children. A radiation-induced mutation in
a body, or somatic, cell has consequences only for the person irradiated, including the child in the womb, since somatic cells do
not enter into the reproductive process. Germ, or sex, cells, on the
other hand, are directly responsible for reproduction, so only if
they are harmed can we pass the damage — whether it be cleft
palate, deformed limbs, or a predisposition to cancer — on to our
children, and they in turn pass it on to their children.)

It doesn't necessarily take huge amounts of radiation to produce
such damage either. Diagnostic x rays, though not considered
high-risk in terms of hereditary harm, are nonetheless known to
cause genetic defects. In 1979 the government commission on the
Biological Effects of Ionizing Radiation concluded that if all
Americans of childbearing age were to receive 1000 millirems, the
equivalent of a fluoroscopic upper gastrointestinal examination,
from five to seventy-five babies out of every million born each year
would have severe radiation-induced genetic disorders. The same
dose received by subsequent generations would result in an increase of 60 to 1100 genetic disorders. Though these are small
numbers compared with the 197,000 severe genetic abnormalities
from natural and environmental factors for every million babies
born annually, it is essential to emphasize the importance of
shielding our sex glands whenever possible while being x-rayed.
And, it is just as important to realize that for many of the most
common diagnostic radiation tests, and forms of radiotherapy for
cancer, the same procedures can deliver significantly more radiation to you than to your mate.

Why Are Men and Women Affected Differently?

Because of the simple anatomical fact that the gonads in men and
women are located in different parts of the body. For some medical procedures, such as dental x rays where the body can be protected from the neck down by a lead-rubber apron, both men and
women receive the same gonadal dose — presumably zero. But in

most x-ray exams this is not the case. For instance, a chest x ray delivers almost double the amount of radiation to a woman's gonads because her ovaries are located closer to the target of the examination. This "sex" difference is particularly marked for women who undergo upper gastointestinal x-ray series which expose the stomach and small intestines to radiation. Her ovaries receive roughly four times as much radiation as the testes of a man undergoing the same test. Unfortunately, there is no way to completely shield the ovaries in such a procedure.

On the other hand, men's testes are subjected to higher doses of radiation from a lower gastrointestinal x-ray series. The barium enema test to examine the large intestines exposes a man to almost twice the gonadal dose a woman receives. The sex difference is even more pronounced with lumbar spine x rays, which expose a woman's sex glands to 275 millirems and a man's to a whopping 2268 millirems — more than eight times as much radiation.

Lumbar spine x rays have been and, in some companies, still are a routine medical procedure before a construction worker or other laborer is hired. Their purpose is to screen men for potential lower-back weaknesses that could be aggravated by heavy work. But since most laborers are physically fit, and often of an age to father children, the indiscriminate use of lumbar spine x rays poses a health risk to the man x-rayed, and, because his testes aren't protected, to his future offspring. If you're interested in learning of the different radiation doses to a man's and woman's gonads from the same diagnostic x-ray tests, you can consult Appendix A on page 203. In many cases, however, we can avoid or decrease the risk of radiation-induced genetic mutations and their potential for birth defects if we take into consideration certain variables like the "time factor."

How Does the Time Factor Act as a Safety Measure?

By allowing a sufficient amount of time to pass between an x-ray procedure that exposes your gonads to a significant dose of radiation and conception, you can greatly minimize the possibility of germ cell damage. This is because damaged chromosomes and genes may heal themselves, given sufficient time, and because in men there is a relatively fast turnover in the elimination of older sperm cells and the production of new ones.)

Radiation biologists suggest that if you are a male and receive a moderate dose of radiation — typically about 10,000 millirems from a nuclear accident or from radiation therapy — you can significantly increase the chance of fathering a healthy child by deferring conception for a minimum of two months, but preferably six months. This does not imply that radiation damage to a man's gonads can always repair itself. If his cells that produce the sperm are harmed, he could become completely sterile — which is unlikely from diagnostic x-ray doses — or produce defective sperm cells. A whopping dose of 250,000 millirems could cause temporary sterility for one to two years, and 600,000 millirems induces permanent sterility.

Since a man's gonads are actively producing mature sperm cells from puberty to his death, they are considered at least twice as vulnerable to radiation damage as a female's ovaries, which do not undergo similar biological activity. The 1979 BEIR committee concluded that men are far more likely than women to pass on genetic defects to their children. As a measure of safety, if a man receives such diagnostic x-ray tests as a barium enema, lumbar spine, or hip examination, he would be wise to defer conception for the suggested two to six months — even though these diagnostic procedures expose his gonads to only about a tenth the dose he might get from radiotherapy or in a serious occupational accident.

How Does the Time Factor Apply to Women?

For a woman the threat of radiation to her sex cells presents a somewhat different problem, since all the eggs she will ever produce are present in her body three days after she is born. Therefore, whatever eggs may be irreparably damaged by radiation exposure will remain damaged throughout her reproductive life so the potential to produce a deformed child is not improbable. Since damage is not always irreparable, a woman who receives a significant dose of radiation to her ovaries would be wise to wait six months to one year before she tries to conceive. If a woman's gonads receive a dose of 400,000 millirems of radiation — an exposure not likely from diagnostic x rays — she will become permanently sterile.

Time isn't the only consideration, though, in determining potential sex cell damage. In addition, there are also "dose" and "radiation" factors that influence our risk.

How Much and What Kind of Radiation We Get Influences Our Health?

Exactly. /A large one-time dose of ionizing radiation produces more mutations of all kinds than the same amount administered in smaller doses over an extended period. In radiation cancer therapy, for example, a patient often receives hundreds of rems at the site of a tumor, but great effort is made to spread such high doses over a period of weeks or months. Since delivering a specific amount of radiation over too long a period of time can dilute the effect of treatment, a radiologist must decide upon the optimum therapy that will kill the tumor and at the same time minimize genetic, as well as somatic, risks to the patient./

Of course the type of radiation you receive will depend on the nature of your medical problem. In therapeutic radiation treatments for cancer, for example, cancer of the bladder usually is treated with x rays because they are an effective form of radiation to penetrate the soft tissue of the lower abdomen and kill malignant bladder cells. But a brain tumor cannot be treated with the same kind of x rays. For one thing, they aren't powerful enough to sufficiently penetrate the skull. Instead, neutron irradiation is often used in the treatment of brain tumors. It is up to the radiologist to select the safest, most effective kind of radiation procedure for a specific problem, whether it be a diagnostic test or for therapeutic purposes.

/By calculating the time, dose, and radiation factors, radiation biologists are able to estimate how much and what kind of radiation is likely to result in our producing children with birth defects. One popular, convenient method of measurement is the "Doubling Dose" — the amount of radiation required to double nature's own mutation rate./Experiments with mice have shown, for instance, that the dose of x rays required to double the natural mutation rate ranges from 10,000 to 200,000 millirems, depending on the type of mutation produced. /

What's the Doubling Dose for Humans?

Most radiation experts believe the doubling dose for men and women to be as low as 30,000 millirems, others claim it is closer to 100,000 millirems. The reason for this wide range is understandable considering that no genetic radiation experiments have ever been performed on humans, and that the total number of people ever exposed to such doses is statistically incredibly small. Accepting this range, the doubling dose indicates that a child born to a parent who has accumulated 30,000 (or 100,000) millirems of radiation has twice the chance to develop genetic abnormalities than he or she would have if the parent had not been exposed. In the unlikely event that both parents accumulate a doubling dose, their offsprings' probability for genetic defects is at least quadruple the natural mutation rate in the human population.

But Aren't My Chances of Receiving a Doubling Dose Slim?

Very slim — barring nuclear war or a serious power-plant accident. This is true even if you are a radiologist, an x-ray technician, or a nuclear-power-plant employee who routinely works around radiation sources.

The doubling dose represents a large amount of radiation. Compared to it, the actual doses received by most of us are extremely small. For example, radiation from natural sources, such as radioactive elements in the earth and air, and cosmic rays in space, expose us to an annual dose of roughly 100 millirems of ionizing radiation. Medical diagnostic x rays, the single largest source of manmade radiation, amount to an average 20 millirems a year for each of us — one sixth-hundreth of the lowest dose estimate required to double the natural mutation rate. Still, it is important to remember that much smaller amounts of radiation than the doubling dose can cause genetic defects.

In certain medical procedures, however, a person can safely receive even higher amounts of radiation than the doubling dose, though exposure is less risky since it's not directed at the sex organs. For instance, in the treatment of an overactive thyroid, radioactive isotopes can deliver as much as 10,000,000 millirems to the thyroid gland.

What's a Radioactive Isotope?

A radiation-emitting element that is either swallowed or injected into the blood stream to diagnose and treat certain diseases. For instance, radioactive iodine-131 is commonly employed to diagnose thyroid problems. In a typical thyroid function test, you would swallow a solution of iodine-131, which concentrates its radioactivity in the thyroid gland. The gland receives a radiation dose of roughly 500 millirems — five times our annual background exposure — while the dose to the rest of the body, in particular the sex glands, is less than ten millirems. Using a sensitive scanning device, your doctor can measure how the thyroid gland is metabolizing iodine, and thus diagnose its state of health.

Are Medical Isotopes Potentially More Dangerous than X Rays?

They can be. This is primarily because the radioactivity from an x ray is a quick, one-shot blast, while radioisotopes continue to irradiate the body for days or weeks. As with x rays, therapeutic doses of medical isotopes deliver significantly more radiation to a patient than do diagnostic doses.

For instance, for the majority of patients diagnosed as suffering from overactive thyroids, the standard treatment is a dose of radiation to the thyroid that is one thousand times larger than the dose used to test the gland in the first place. Most of the radioactive iodine-131 concentrates in the thyroid, subjecting the gland to intense irradiation and reducing its activity to a normal level within a few days. As pointed out, the radiation dose to the thyroid may run as high as 10,000,000 millirems, and the average dose to the patient's entire body, from radioactive iodine in the circulating blood, is roughly 14,000 millirems. These are large amounts of radiation and can be justified only if there are substantial health benefits to offset the risks.

/While eight out of ten thyroid patients treated with radioactive iodine are cured, a small percentage of them, as well as patients treated with other radioisotopes, develop leukemia and other forms of cancer, and risk parenting children with gross abnormalities./

/A pregnant woman who must undergo a medical isotope test should be aware of an additional hazard: certain radioisotopes cross through the placenta and enter the developing fetus, where they can easily produce mental and physical abnormalities./ If you're interested in finding out the radiation doses administered in many common isotope tests, consult Appendix E, page 207. The doses to fetuses of mothers who submit to isotope tests are given in Appendix F, page 208.

Reducing the Risks

Since X Rays Can Be So Harmful, Why Are So Many Unnecessary Ones Taken?

The underlying cause of x-ray abuse is a pervasive belief that most diagnostic x rays expose us only to low levels of radiation that are very safe. After all, no conscientious doctor would routinely x-ray a patient annually or semiannually if he or she believed the radiation was seriously harmful.

For years many radiation biologists and doctors have adhered to the so-called threshold theory regarding all forms of ionizing radiation. According to this theory, the risk of cell damage resulting in bodily harm drops to practically zero below a certain threshold of radiation — typically anywhere from several hundred to a few thousand millirems accumulated over a year. Being soft and low-level, most diagnostic x rays fall below this threshold and, therefore, have been assumed to be practically risk free.

Not all radiobiologists, though, agree with the threshold argument. Early on, a relatively small group of them suspected that radiation damage occurred in a linear fashion directly proportional to dose, with no dramatic cutoff point at any level. According to this "linear theory," the only time there is no adverse biological effect, or cell damage, is when there is no dose; granted, at the very lowest doses the risk for bodily harm may be practically nil. The linear theory has been attacked by threshold advocates, who claim that it exaggerates the risk of low-level radiation, generally regarded as an annual exposure of less than 10,000 millirems, or 100 times greater than the dose we get from natural background radiation.

How Much Radiation Can We Safely Absorb?

There is no clear-cut answer. The scientific community is split over the danger of low-level radiation because its effects are subtle, can require decades to show up, and can easily be masked by other injuries to the body. Nevertheless, an increasing number of scientists are coming to the conclusion that there is no threshold below which ionizing radiation is safe. In fact, in their 1979 report, a majority of the members of the National Academy of Sciences' Committee on the Biological Effects of Ionizing Radiation reiterated their earlier support for the linear theory, concluding that no radiation is without some risk. The majority conceded that, while the linear theory "probably overestimates" the risk from less harsh, sparsely ionizing radiation like x rays and gamma rays, the risk was real enough to embrace the theory as a matter of prudence. On the other hand, for highly ionizing radiation like alpha particles and neutrons — the kind we could get from medical procedures, or A-bomb blasts, or a nuclear-power-plant meltdown — the majority feel that the linear hypothesis is probably more likely to underestimate than overestimate the risk.

In short, while lower levels of radiation are being viewed with increasing suspicion, the earlier widespread belief that they were "safe," and the persistent belief among many doctors that they still are, have directly contributed to what may be termed the more immediate causes of the misuse of x rays.

What Are the Other Causes of X-Ray Abuse?

First, as was pointed out earlier, many doctors are becoming increasingly afraid of malpractice suits. To cover themselves they often recommend complete x-ray examinations, even when they don't seriously suspect that a patient is all that ill. In this respect, the x-ray films can protect a doctor in the event that a patient later develops a medical problem he or she claims the doctor over-looked. Unfortunately, as a result, many doctors today draw little distinction between defensive medicine, practiced with one eye on the patient and one on the lawyer, and preventive medicine, where the need of x rays can be very real. One 1973 study by researchers at the Massachusetts Institute of Technology estimated that 30 percent of all x-ray exams are conducted for malpractice protection.

Another reason for the dramatic rise and abuse of x rays has been a change in emphasis in health care. Once x rays were used mainly to diagnose disease. Now they are also used to monitor complications that may result from a disease, to chart a patient's course to recovery, and for routine checkups. When utilized in this way they can serve as a reliable tool, but the emphasis on preventive medicine has resulted in millions of us receiving unnecessary radiation from routine chest x rays for hospital admission, periodic lower spine x rays for entry into some heavy labor jobs, twice yearly dental x-ray checkups, and assorted diagnostic x rays that have become a standard feature of what is referred to as "executive employees physicals," offered by many companies. There is often no need for such wholesale scheduling of x rays for people with no apparent symptoms of disease.

In addition, many of us receive excessive doses of x rays simply by being examined by physicians who are not trained in radiology. It is an alarming fact that every doctor and dentist is licensed to operate an x-ray machine, even though his or her training in radiology was probably limited to a few lectures and demonstrations in medical school. The problem is compounded by the fact that many private physicians owning x-ray devices have not invested in the latest safety equipment — or are unfamiliar with improved irradiation procedures. In fact, because of carelessness and poor filming techniques, x rays often have to be retaken.

How Often?

Enough to make reshot x rays an important health concern. Recent government estimates are that 5 to 10 percent of all diagnostic x rays must be repeated because of incompetence or carelessness. This means that at least a quarter of a million x rays must be reshot every year, exposing a person on the average to twice the amount of radiation he or she would have received. This is even more disturbing considering that in many cases the x rays may have been unnecessary in the first place. In one study more than half the hospitals monitored by the Food and Drug Administration reshot x rays of their patients because the first pictures were taken incorrectly.

An even larger problem involves radiation equipment. According to the Bureau of Radiological Health, as of 1979 about 30 percent of all x-ray machines failed to comply with one or more of the federal exposure safety tests. Though the bureau has set standards, compliance is largely voluntary. A recent government survey of 1000 x-ray machines showed that they were delivering up to one hundred times more radiation than necessary for an average chest x ray. Another study of 35,224 dental x-ray units in thirty-eight states revealed that 36 percent of them emitted "excessively high" doses of radiation. What's more, fewer than ten states require x-ray technicians to be licensed.

The National Academy of Sciences has estimated that if all of the unnecessary or repetitive x-ray examinations were eliminated and x-ray technology improved where needed, the number of diagnostic x rays could be reduced by at least 30 percent.

Is Anything Being Done to Cut Back on Our Exposure?

Yes, there has been some movement to reduce our exposure to unnecessary x-ray radiation. For example, because of a recent presidential directive dental x rays can no longer be routinely taken as screening tests in government hospitals. Instead, they must be justified by a dentist. Some insurance companies are exploring the possibility of insisting on similar justification before reimbursing policyholders for dental x rays. And an increasing number of states are considering licensing x-ray technicians and imposing stiff monetary penalties on all x-ray machine owners who violate manufacturer's safety standards.

While these and similar attempts to cut back on radiation exposure are important, in terms of overall x-ray reduction they merely scratch the surface. What is really needed to drastically cut our unnecessary exposure to x rays is a concerted effort by government, doctors, and radiologists to stop the casual administering of them. For those of us, however, who have to have x rays for either diagnostic or therapeutic purposes, new medical procedures are greatly reducing radiation doses. There are new devices — some in the experimental stage — being utilized in a number of hospitals and medical centers across the country that can significantly diminish or eliminate radiation exposure by either complementing existing medical techniques or replacing them entirely, so we can protect ourselves, when possible, from needless radiation.

What Kind of New Techniques?

There are improved x-ray films and devices known as intensifying screens that are providing doctors with high quality images of bones and internal organs while delivering much less radiation to the patient. In the early days of x-ray radiography ordinary photographic film was used to take x rays. Gradually, though, over the years improved films were made and today they are being coated with chemical emulsions to greatly enhance their sensitivity to x rays. The more sensitive a film, of course, the smaller the dose of radiation needed to produce an image. Some of the newer sophisticated films are not yet widely used, while other more sensitive ones are still in the experimental stages. However, it won't be long before most of us will be benefiting from the more sophisticated film.

Intensifying screens can similarly reduce our radiation exposure. These thin sheets of plastic, or cardboard, are coated with a substance that fluoresces or emits blue light when struck by x rays. The fluorescence acts with the x rays to produce a picture of a bone or internal organ and thus reduces the amount of radiation needed. Some of the most sensitive intensifying screens are being made from rare-earth metals, which substantially reduce the dose of radiation required in many diagnostic procedures.

In addition to newer x-ray films and intensifying screens, scientists have found that by passing electrons through silicon crystals they can produce incredibly well focused beams of x rays. The technique was discovered in 1979 by scientists at Stanford University and the Lawrence Livermore Laboratory in California. It shouldn't be too long before such finely tuned beams are used for diagnostic as well as therapeutic purposes, in which they could kill tumors with minimum damage to surrounding healthy tissue.

One of the biggest diagnostic x-ray breakthroughs, though, is the CAT scanner. CAT scanners have been in use since the mid 1970s, and doctors have been able to perform precise examinations of the brain, and more recently the body, by using the CAT, short for computerized axial tomography. So important is this revolutionary x-ray scanning technique that its developers, Godfrey Hounsfield and Allan Cormack, won the 1979 Nobel Prize in Physiology and Medicine.

How Does the CAT Work?

A CAT scanner consists of an x-ray tube and an electronic detector positioned on a circular track. When the devices are set in motion they rotate opposite each other around a patient's body, snapping thousands of x-ray pictures in a matter of minutes. The snapshots are processed by a computer which creates a cross-sectional view of the patient's body on a video screen. A visual slice can be taken of any portion of the body and many closely spaced slices can be combined to form a complete picture of an organ. With such information stored in a computer, a doctor can choose to view an organ from any perspective, slice it through any plane, or activate a zoom lens to examine its particular features close-up.

You might think that the x rays required to produce such an organ composite would expose a person to more radiation than a conventional set of x rays. But that's not always the case, because a CAT scanner produces such clear precise pictures that the procedure seldom has to be repeated. Plus, a single CAT scan can also reduce a number of painful and potentially hazardous tests in which chemical dyes are injected into arteries, or air into the brain, so that conventional x rays can detect tumors, birth defects, and strokes. There is another type of scanner called PET, an acronym for positron emission tomography. It is still in an experimental stage but when in wide use will also substantially cut back on radiation exposure in medical procedures.

How Does PET Differ from CAT?

A PET scanner is a device that detects radiation emitted from a patient's body after he or she has swallowed, inhaled, or been injected with a drug tagged with a radioactive element called a radioisotope. In most conventional medical scanning procedures of this kind, the radioisotope retains a substantial portion of its radioactivity for several days, continuously irradiating the patient's body. With PET scanners, however, drugs are tagged with radioactive elements that shed most of their radiation in just a few minutes or hours in the form of subatomic particles called positrons. Since these radioisotopes possess such short half-lives, drugs tagged with them must be prepared shortly before they are administered if they are to be effective. This means that scientists working with PET scanners must have ready access to high-energy accelerators that produce the positron-emitting isotopes.

So far, researchers at six centers in this country have conducted PET studies by tagging numerous substances, including alcohol, with the fast-decaying positrons, as well as sugar, amino acids, fats, and proteins. The research is being backed by a $10 million grant from the National Institutes of Health, and if all goes according to plan, PET scanners should be a practical medical tool by the end of this decade. When in operation, PET scanners will allow doctors to make more accurate diagnoses, and treat a wide range of diseases, with remarkably less radiation than is required today.

Another breakthrough in the use of radioisotopes, called radioimmunoassay, is already being employed to reduce our exposure to radiation. Also a Nobel Prize–winning technique, it was developed at the Bronx Veterans Administration Hospital by Dr. Rosalyn S. Yalow and the late Dr. Solomon Berson, and allows doctors to use radiation to search for and diagnose certain diseases without the radiation touching the patient.

How Does Dr. Yalow's Technique Work?

A doctor simply takes a tissue sample or fluid from a patient, then uses radioisotopes to perform tests in the laboratory rather than in the body. In the late 1970s about sixty million of these tests were conducted yearly for a wide variety of purposes ranging from screening blood transfusions for hepatitis to diagnosing hormonal disorders. The technique has become so popular that its use is expected to double by 1985.

Other new medical procedures are actually replacing radioactive isotopes by employing their nonradioactive chemical cousins. For instance, carbon-14, widely used for many diagnostic tests, has intense, long-lived radioactivity that prevents it from being used in infants, children, and pregnant women — except as a last resort. But its cousin, carbon-13, is nonradioactive and safe. Scientists at the Argonne National Laboratory in Illinois have recently devised a procedure employing carbon-13 that is believed to have great potential for diagnosing gastrointestinal disorders.

With the technique, a patient ingests a drug or food in which the atoms of carbon-12 — the common, widespread variety — have been replaced with carbon-13. The body's metabolic processes then convert the carbon to gaseous carbon dioxide, which is exhaled. By sampling the patient's breath and examining it for the presence of carbon-13, doctors can determine if the stomach, liver, pancreas, and several other organs are functioning properly. So far the tests have been successful in a year-long clinical study, and in procedures where carbon-13 can be used it reduces a patient's radiation dose to the only truly safe level, zero.

There is also no radiation from another diagnostic technique, ultrasonography, that relies on sound to examine internal organs or a fetus in the womb. The process has already replaced many diagnostic x-ray procedures, especially those conducted on pregnant women.

How Does Ultrasonography Operate?

Ultrasonography employs high-frequency sound waves that are beamed at internal organs and ricocheted back to form fairly detailed images on video screens. The process itself is not new. Obstetricians have used ultrasound since the 1960s to examine fetuses for abnormalities, because it poses no radiation threat and can detect features ten times smaller than those located by x rays. But newer ultrasonic machines that provide even greater clarity are expected to revolutionize the field of ultrasonographic medicine in this decade, substantially reducing medical reliance on radiation.

Can Radiation Exposure Be Reduced in X-Ray Therapy as Well as in Diagnostic Tests?

There have been significant achievements in this area as well. While x rays and gamma rays still are the most widely used forms of radiation in treating cancer patients, other, particle radiations like neutrons and protons have shown promising results in clinical tests. Since they kill tumors more effectively, smaller doses can be used. These are still largely experimental treatments, but they look promising for the future. Perhaps one of the most auspicious prospects for reducing radiation in therapeutic treatments is pion therapy.

What Are Pions?

Pions are subatomic particles that occur abundantly in nature and are believed to be the glue that holds together the nuclei of atoms. When used in the treatment of cancer, pions can greatly reduce the amount of radiation the patient receives.

The largest doses of radiation we can receive from medical procedures are those administered to destroy tumors. Unfortunately, when a cancer victim receives such huge doses of x rays or gamma rays they can widely scatter, often killing more healthy cells than cancerous tissue. This is not the case with pion therapy. A beam of pions can be driven to a precise depth and location within a patient's body where they explode into "pion stars." These showers of nuclear radiation fly in every direction, but only over very short distances, sparing healthy tissue while basically killing that which is diseased.

Pion beams were successfully tested on cancer patients in the late 1970s at the University of New Mexico with highly encouraging results. In many patients tumors shrank and remained in remission for almost a year, and virtually all patients experienced less of the vomiting and diarrhea that accompanies conventional radiotherapy. From these tests, doctors determined that pions appear to be almost 40 percent more effective than x rays in killing cancer cells. Clinical studies are now being conducted at research centers in this country, Switzerland, and Canada, and the treatment could be available to many cancer patients before the end of this decade.

The supereffectiveness of pions, and to a somewhat lesser extent, neutrons and protons, is even more encouraging when you consider that they'll eventually be used in conjunction with new drugs that increase the sensitivity of cancerous tissue.

What Are Some of These Drugs?

Scientists are experimenting with two kinds. The first class of drugs consists of "radiation protectors." As their name implies, when injected into a patient's body prior to radiotherapy they actually diminish the damage done by ionizing radiation. They do this in effect by acting like sponges within the patient's cells to soak up highly reactive particles known as "free radicals" which are produced by irradiation. Thus the radicals are quickly prevented from doing any damage.

The second class of drugs is made up of "radiation sensitizers," and their function is different. They possess the highly desirable property of being able to increase the sensitivity of malignant cells but not healthy cells to radiation injury. Though their exact mode of action remains unknown, it's believed that the members of at least one group of sensitizers act by insinuating themselves into a malignant cell's DNA chain, which weakens the chain and makes it more susceptible to radiation damage. Another group of sensitizers is thought to work by changing the chemical environment of malignant cells — actually by increasing their oxygen content — so that when radiation damage occurs, the cell cannot heal itself. The damage becomes "fixed," or permanent.

Within the last decade, several protectors and sensitizers have been tested on cancer patients undergoing radiotherapy with x rays, as well as with neutrons and protons. Unfortunately, several of the best drugs in both categories are highly toxic in the large quantities in which a patient must take them to gain their advantages. Even in smaller amounts, where their effectiveness is slight but real, the drugs often produce such unpleasant side effects as nausea, vomiting, and dizziness.

However, there are several drugs of both kinds that hold hope for the future. The manner in which these drugs must be administered to the patient in general, or to the specific organ to be irradiated, makes their use at the present time highly complex, limited to certain kinds of cancers, and still experimental. Currently they are used only at the best medical centers, under the supervision of a radiologist, pharmacologist, and physician. A patient who suffers from a severe malignancy that has not responded well to radiation therapy alone, should ask his or her doctor about the possibility of employing radiotherapy with drugs. In the event that drugs are actually used, the patient should be informed about any side effects or toxicity.

II Nature's Own Radiation

Our Natural Background Exposure

Is Radiation from Natural Sources Similar to X Rays?

Like x rays, natural radiation, also known as background radiation, is the ionizing kind that damages cells by ripping electrons off atoms. Natural radiation is all around us, in our soil, air, even in our bodies. Each of us receives an average annual dose of roughly 100 millirems of natural background radiation (NBR) a year.

The largest dose of natural radiation we receive, about 50 millirems a year, comes from soil and air, in the form of invisible gamma rays emitted from rocks and radioactive minerals, and from radioactive elements, such as radon, adrift in the air. The second largest dose, 30 millirems a year, originates from cosmic rays emitted by the sun and stars within our solar system, and from other galaxies. Cosmic rays are high-energy radiation particles that constantly bombard the earth from outer space and pass through our bodies.

Radiation from cosmic rays, soil, and air constitute what is called external background radiation. But there is another source of NBR that is internal. It comes from background emissions continuously released by traces of radioactive elements in our body tissue. The amount of internal radiation we absorb depends on our diet, and the amount of radioactivity in foods can vary dramatically. Nuts and cereals, for example, contain higher levels of radium than milk or meats, and the radioactive content of drinking water around the world varies by a factor of 10,000. Internal radiation, however, constitutes the smallest background dose we receive — only 20 millirems a year.

How Harmful Is Natural Radiation?

All radiation is potentially dangerous to life. Although the exact percentages are unknown, natural background radiation is thought to be responsible for a small portion of all cancers and genetic disorders afflicting us today.

Fortunately for most of us, though, the unavoidable NBR we accumulate is too insignificant to pose a threat. In fact, NBR is believed to have played a major role in the birth and evolution of life on earth, helping to create genetic mutations in plants and animals which make them more (or less) adaptable to their environment, and, therefore, more (or less) capable of survival. Whatever role background radiation has played in our evolution, considering our adaptability and intellectual development, its effects apparently add up to a net plus. Still, there are situations in which we can be exposed to unacceptable, damaging, and even deadly levels of NBR. We should be aware of these levels and understand that some of us may be getting too much exposure for our own good.

How Come Some People Are Exposed to More Natural Radiation than Others?

Because radiation levels in soil and air differ significantly around the world. So, the country, city, or town where we live — as well as the building materials which originate from these areas — can substantially affect the amount of NBR we receive. Also, we are exposed to varying amounts of cosmic radiation from space depending on the latitude and altitude at which we live. Some people reside in geographical "hot spots," areas of the world where the levels of natural background radiation are considerably higher than usual. These locales are "hot" because of radioactive elements that were trapped in cooling lava and condensing water vapor when the earth formed some 4.6 billion years ago. In these hot spots water can be 10,000 times more radioactive than what most of us drink. Rocks mined from these areas and bricks molded from the soil and used for buildings continually irradiate the local populations with possibly detrimental effects.

Where Are the Prominent Hot Spots?

If you live in France, Brazil, Egypt, the Niue Island in the Pacific, the Kerala region of India, and to a lesser extent Sweden, there's a chance you're living in one of the more prominent hot spots.

Alum shale deposits in Sweden expose thousands of people to 160 to 220 millirems of NBR a year — double what most Americans receive. In France, seven million people, about one-sixth of the French population, reside in regions where the surface rocks are principally granite. These rocks emit background radiation yielding doses from 180 to 350 millirems a year — three times the average NBR received by most of the world's population.

In Brazil, the dose to some tens of thousands of people is even higher. Several areas of the country along the coast are rich in radioactive monazite alluvial deposits. The combined dose from the soil and rocks is 500 millirems a year. In a few sections of the country the dose rises to 1000 millirems a year, placing the inhabitants in the same exposure category as x-ray technicians and nuclear-power-plant employees, who are routinely exposed to such radiation levels.

A densely populated area along Egypt's northern Nile Delta contains various radioactive minerals in the soil. In several towns radiation doses of 300 to 400 millirems have been measured.

Residents of Niue Island in the Pacific receive even more. Their natural background radiation comes from a combination of volcanic soil and the unusually high radioactive content of the vegetation that grows in it. The average annual dose to the people of the island is 1000 millirems a year. Inhabitants who consume large quantities of vegetation greatly increase their internal radiation dose.

People who live in the Kerala region of India get much more NBR. The region lies above a bed of intensely radioactive rocks and monazite sands rich in the radioactive element thorium. It is estimated that more than 100,000 people in the region receive an average of 1500 millirems of radiation a year. In certain sections of Kerala the dose reaches a whopping 3000 millirems a year — twice the exposure a woman's breast receives in a mammogram or breast x ray. These are the highest levels of natural background radiation so far recorded on earth.

Do People Who Live in These Hot Spots Suffer Any Ill Effects?

There is growing, though inconclusive, evidence that they do, and ongoing studies may help us understand more about the effects of natural, as well as manmade, radiation exposure, even at their lowest levels.

In 1976 four physicians from the Department of Medicine of the All-India Institute of Medical Science studied the inhabitants of a coastal area of Kerala to determine what effects the continuous exposure to 1500 to 3000 millirems of radiation a year might have. The doctors compared the health of the Keralans with that of residents of a village several miles away who share the same life style and diet, but who experience an annual background dose of only 100 millirems — what most of us receive each year. The doctors found that in the Kerala group the incidence of mongolism, or Down's Syndrome, and other forms of severe mental retardation of genetic origin was four times higher than in the neighboring villages. Most of the retarded children had been born to mothers between the ages of thirty and thirty-nine, suggesting that the natural risk of older women delivering abnormal children may be increased by exposure to low levels of radiation. In Brazil, investigators studying a group of people living in an area where annual NBR levels reach 1000 millirems have detected chromosome damage that can lead to birth defects.

These annual levels are well below the Maximum Permissible Dose (MPD) of 5000 millirems a year that every American *radiation worker* is allowed to accumulate according to federal safety standards. Yet until more convincing evidence of the effects of NBR surfaces, scientists caution about reading too much into these preliminary findings.

Why?

For one thing, the studies have focused on small segments of population that are themselves very tiny; the total number of people who live in the hottest of hot spots is minuscule compared to the world population. There is also the complex problem of trying to determine the effects of NBR on peoples of differing races and nationalities, who possess diverse work and living habits as well as varying amounts of stress. These variables can seriously complicate any medical investigation. What's more, scientists have yet to detect any increase in the incidence of cancer, particularly leukemia, in Keralans or inhabitants of Niue Island or any other hot-spot populations. Leukemia is usually one of the first forms of cancer to appear as a result of irradiation.

Critics of the Kerala study point out that, even if small differences in the incidence of cancer were found among people who live in radiation hot spots, the increase could be attributed to other causes, particularly the fact that Keralans occasionally marry close blood relatives like first cousins. Even if natural radiation in these hot spots is proven to be relatively harmless, we should keep in mind that in addition to this natural radiation there are increasing amounts of manmade radiation in our environment, and there is growing concern — and evidence — that the indiscriminate use of dental and medical x rays, plus radiation from a host of other modern gadgets, are slowly and insidiously undermining our health. If you live in a hot spot you start out with much higher average natural levels to begin with, and manmade doses then pose an additional threat.

Are There Any Hot Spots in This Country?

While we have no known geographical hot spots comparable to Kerala and the other prominent locations, we do possess a large "cosmic" hot spot in the Colorado Plateau. There are numerous cosmic hot spots around the world and they are basically equivalent to the geographical ones, only the radiation emanates from cosmic rays in space rather than from within the earth.

Cosmic rays consist of potentially hazardous protons, helium atoms, the nuclei of heavier atoms, and some electrons. They interact with our atmosphere to produce secondary radiation in the form of electrons, gamma rays, and mesons, a class of subatomic particles. If it weren't for the protective effect of our atmosphere this radiation would annihilate us. Still, some cosmic rays seep through, and the exposure we receive depends on the latitude and altitude at which we live.

Cosmic radiation levels differ with latitude because of the earth's magnetic field, which deflects cosmic particles away from the equator and funnels them into the polar regions. So if a person lives in the tropics, he receives less cosmic radiation than if he lives farther to the north or south. For instance, if you live in a city near the zero-degree latitude of the equator, you receive about 35 millirems a year; if you live at 50 degrees latitude, in London, Seattle, or Moscow, you receive about 50 millirems a year. But this difference due to latitude is slight. What substantially increases the dose of cosmic radiation we receive is altitude.

Why Is Altitude Crucial to Radiation Exposure?

Because the higher up we go the thinner the air. Thinner air contains fewer water molecules to absorb cosmic radiation. As a result, for every 5000 feet we ascend above sea level, the intensity of cosmic ray bombardment approximately doubles. If you live in the high plateaus of Wyoming, for instance, you get an average annual dose of 115 millirems of cosmic radiation compared with only 45 millirems if you live in the lower plains of Kentucky and Indiana. Residents of the "mile high" Colorado Plateau reside in a cosmic hot spot; they accumulate 180 millirems of cosmic radiation a year.

Variations in cosmic radiation can be drastic. Some of the most populated areas of the earth, like the Tibetan Shelf that includes parts of China, are at altitudes close to 15,000 feet, where cosmic ray doses run as high as 300 millirems a year. At the top of Mount Everest, the highest point on the earth's surface, cosmic radiation doses reach 800 millirems a year. Most of these levels fall well within what is currently considered to be the safety limit of 500 millirems to our entire body in a year. No one, though, can be ab-

just what subtle damage even these very low levels
on our body cells when combined with the increasing
manmade radiation in our environment. What we do
however, is that we have to ascend much higher to experi-
oses that are known to damage our bodies.

at Kind of Damage?

rain damage for instance. NASA scientists were surprised to
learn that during the Apollo 11 trip to the moon one of the astronauts reported "seeing" bursts of light while resting with his eyes
closed in a completely darkened cabin. Almost all of the crew
members of later lunar missions reported experiencing similar
light flashes at the rate of one or two a minute. Some described
them as shooting stars, others as hazy luminous clouds. Scientists
suspected cosmic radiation was responsible, and recent laboratory
experiments on volunteers have confirmed that the bursts are due
to high-energy cosmic particles that pass through the eye and
collide with the sensitive light-detecting cells of the retina. The radiation, called HZE particles, possesses such enormous energy
that it easily penetrates the walls of a spacecraft and passes
through the body. For example, on a two-week round trip to the
moon about 100 of these cosmic particles will shoot through an
astronaut's head, and as each one pierces the brain it destroys
cells, or neurons, located along its path.

Have Any Astronauts Been Harmed?

So far astronauts who have been assigned to lunar expeditions
show no apparent ill effects from the particles. NASA doctors,
however, are concerned that increased bombardment resulting
from longer space journeys, lasting up to several months or years,
might impair an astronaut's mental capabilities or nervous system,
so they are attempting to determine the hazards of prolonged
HZE exposure. One particularly disturbing fact is the nature of
HZE destruction. All of us continually lose brain cells as we age,
but we have so many billion cells and lose them so randomly that
we are hardly affected by the loss. While the number of brain

cells killed by HZE particles during space flight may be substantially less than our natural loss, the cells are destroyed in contiguous rows. You can imagine the pattern of damage by picturing a line of dead or heavily burned brain cells along the tracks of the HZE particles surrounded by a column of injured or lightly charred brain cells. The effect is similar to passing a hundred thin, white-hot needles through the brain. The destruction is potentially more serious, since neurons are among the few body cells that do not reproduce after childhood.

Does Anything Like This Happen When I Travel by Plane?

No, because at lower altitudes the atmosphere offers sufficient protection to shield us from HZE particles. However, this doesn't mean our exposure to cosmic radiation is always negligible in a plane. By climbing to an altitude of 30,000 to 40,000 feet in a jumbo jet, or 50,000 to 60,000 feet in a supersonic jet, we shed a good deal of the protective shield provided by the earth's atmosphere. On a commercial trans-Atlantic flight from New York to London you get a cosmic radiation dose of about 5 millirems. This is a small exposure but it can add up. A business executive or diplomat who makes four round trips a year across the Atlantic accumulates an additional 40 millirems, more than a quarter of his or her natural background dose. This is approximately equal to the difference in natural background radiation you would receive if you moved from New York to Colorado.

The levels that most of us get from airline travel are generally considered to be safe, but pilots and crews of jetliners who make regular flights or fly polar routes across the hemispheres can accumulate much larger doses that could prove harmful. For safety reasons, that have nothing to do with radiation exposure, pilots and flight engineers can fly only 1000 hours a year, and during this time it is easy for them to accumulate 1000 millirems of radiation — double the annual safety exposure allowed to the American public. In terms of occupational radiation standards, this places pilots and crews in the same category as radiation workers. For stewards and stewardesses, whose flying hours aren't limited, the radiation dose from cosmic rays can be much higher. And on supersonic transport planes, like the Concorde, the potential dangers for crew as well as passengers can be even greater.

Should I Avoid the Concorde?

No, but if you fly in a supersonic plane, you should be aware of the small additional cosmic radiation exposure you accumulate due to increased altitude and the possibility of much greater doses from solar storms.

In order to operate efficiently supersonic planes must fly at least 10,000 to 20,000 feet higher than other aircraft. At that height, the level of cosmic radiation is about four times more intense than at altitudes flown by conventional jumbo jets. However, because supersonic planes halve flying time, in most cases, the cosmic radiation dose a person receives is only 5 millirems, the same as if he or she flew a jumbo jet. While the future of supersonic travel may be in doubt, because of economic and environmental factors, there are supersonic planes on the drawing boards that, if built, will be able to travel faster, and, therefore, have to fly at even higher altitudes, further increasing the risk of radiation. For example, the plans for America's Supersonic Transport, which was never produced, called for the plane to fly at 80,000 feet, where cosmic ray intensities are eight times higher than at altitudes traveled by conventional planes.

The greatest potential for dangerous doses of cosmic radiation to airline passengers and personnel, as well as spacecraft crews, comes from solar storms. These storms produce intense sunspots and solar flares that greatly increase the amount of cosmic particles in space. In fact, the radiation can be so intense that it disrupts radio communications on earth and increases the incidence of skin cancer. The solar-storm risk factor was taken into consideration by the designers of the Anglo-French Concorde. On board the supersonic jet is a radiation detection device to alert the pilot if radiation in the cabin reaches the danger level — a threshold of 50 millirems an hour. This may not seem like much radiation, but on a 3½-hour New York–London flight, a 50-millirem dose per hour would expose you to more background radiation than you receive in a whole year on the ground. Should a pilot detect this hazardous rate, he or she is instructed to descend to a lower altitude.

What's the Chance of a Solar Storm Occurring?

Solar storms roughly occur in 11-year cycles, and in recent years vigorous storms have appeared just about on schedule, in 1948, 1959, 1970, and, more recently, in 1980. But solar storms can pop up at anytime, and this unpredictability has been a serious concern of space flight planners. In fact, it was unexpected solar activity that brought Skylab crashing to earth in Australia in 1979. Solar storms heated up the earth's atmosphere causing it to expand. This expansion placed a drag on the orbiting satellite and forced it down.

The drag effect, though, is only part of the story. A solar storm can deliver a wallop of 100,000 millirems of cosmic radiation an hour to an astronaut — enough to provide a lethal dose in just five hours. No practical amount of lead or concrete can shield astronauts, or anyone else for that matter, from cosmic rays. It was precisely because of the potential dangers presented by solar storms that Apollo moon shots were scheduled to coincide with periods of low solar activity. In the future, as space missions become longer, solar storms are sure to complicate scheduling and prove an even greater problem for fully computerized airplanes and spacecraft.

Could Solar Storms Interfere with Our Space Program?

Possibly, because cosmic rays may cause computers to malfunction. If you have ever used an electronic computer, you probably know that on certain occasions a single digit will change for no apparent reason, or the computer will suddenly shut itself off. Within the computer industry these malfunctions are known as "soft fails" as opposed to "hard fails," which are caused by faulty circuits. In a large computer, soft fails are known to occur about once every 1000 operating hours and can cost a company both time and money. But it wasn't until 1978 that computer engineers discovered that soft fails are often caused by natural background radiation. That was because until the mid 1970s electronic components in computers weren't sensitive enough to be affected by background radiation.

Radiation is mainly responsible for the failures because the substances used to manufacture electronic circuits are contaminated with minute amounts of naturally occurring uranium and thorium. These radioactive elements emit subatomic alpha particles which produce bursts of energy sufficient to interfere with a computer's circuitry. Once scientists realized this, they wondered if other forms of natural background radiation, especially cosmic rays, could produce the same result. In 1979 a group of IBM engineers experimented with the problem. They calculated that cosmic particles can affect computers, and these particles should have the slightest effect at sea level, more of an influence aboard planes, and a significantly greater effect aboard spacecraft.

The results of their investigation could affect each of us directly, since we will be relying on computers with increasing frequency, and their circuits will be even more sensitively designed. Frequently occurring soft fails could have dire implications for a society that depends on computers to operate its lifesaving medical equipment, to perform its complex business transactions, and perhaps most importantly to defend itself against nuclear attack. It will require a leap in scientific ingenuity to solve the problem of computer soft fails from natural background radiation.

Protecting Ourselves from the Sun's Ultraviolet Rays

Isn't Radiation from the Sun Part of Our Natural Background Exposure?

No, it's not. Although it's hard to think of anything more natural and ubiquitous than sunlight, the term *natural background radiation* is reserved only for ionizing radiations such as cosmic rays and x rays. The sun's ultraviolet radiation — only 3 percent of which bathes the earth — is nonionizing radiation which damages our cells basically through chemical reactions that prematurely age skin and can lead to skin cancer. Since the sun's radiation is non-ionizing, "millirems," the units used to measure exposure to ionizing radiation, do not apply.

Fortunately, virtually all of the sun's deadliest ultraviolet (UV) rays, the UVC kind, are absorbed by the ozone layer in the earth's upper atmosphere, and never reach us. These extremely short rays are known to destroy life, and UVC rays produced in the laboratory can kill bacteria and induce genetic mutations in animals. UVC rays can penetrate the ozone layer only after an exceptionally large solar storm. Though the amount of these rays reaching earth is very small, they are suspected of contributing to the increase in skin cancer that occurs two to three years after intense solar activity. There are two other kinds of ultraviolet radiation, however, that seep easily through the ozone shield in significant amounts. They are the medium-length UVB, or burning, rays, and the longer UVA, or tanning, rays.

Tanning and Burning Rays Are Different?

Yes they are, and their strengths vary according to the time of day. UVB radiation is most intense between 10:00 A.M. and 2:00 P.M. Eastern Standard Time or 11:00 A.M. and 3:00 P.M. daylight-saving time. UVB radiation is the cause of painful sunburn, and in large or repeated doses UVB can cause skin cancer. The slightly less energetic UVA radiation predominates in the morning and late afternoon hours, and, depending on your skin type, prudent exposure can safely produce a golden-brown tan.

Under normal conditions UVA radiation is harmless. But if we ingest certain chemicals called photosensitizers, which occur in smog, industrial wastes, and antibiotics, our skin can be burned as badly by UVA rays as by UVB rays. What's more, recent evidence indicates that both kinds of ultraviolet radiation can act upon each other to harm our bodies. This means that if you sunbathe at noon and get a light burn (from UVB rays), then expose yourself to the sun's weaker UVA rays later in the day, you can acquire a severe burn because the UVB rays have sensitized your skin to the weaker UVA light. Repeated burns of this kind, as well as severe burns from exposure to just midday UVB rays, can dangerously damage the skin.

Is Such Sunburn Damage Permanent?

That depends on the extent and duration of your burn. The effects of mild exposure to ultraviolet radiation include a reddening and swelling of your skin due to dilation of the blood vessels, referred to as erythema, the Greek word for flushing. This condition is usually reversible and your symptoms disappear within several days. But repeated heavy doses of ultraviolet radiation from the sun, and sunlamps, too, damage blood vessels that nourish your skin and take weeks to heal. As a result, your skin's elastic fibers degenerate, the surface layers thicken and wrinkle, and eventually the skin becomes leathery and blotchy. In simple terms, we age prematurely. Farmers and laborers who have worked years in the sun are prime examples of how ultraviolet radiation can accelerate the external appearance of aging. If you want to find out what your face would look like if it had never been exposed to the sun, take a look at your buttocks.

A little-appreciated fact is that sun-related skin changes associated with old age can occur from a single summer of repeated sunburns — though they may not manifest themselves for ten or twenty years. Unfortunately, once your skin has become prematurely aged by ultraviolet radiation no creams, vitamins, or diet can return it to its youthful state. Repeated burns from ultraviolet radiation can also contribute to the development of scaly gray lesions, keratoses, and ugly dark splotches called liverspots, that we're forever buying creams to erase. These are often precancerous and in later life turn into malignant growths.

Ultraviolet radiation can act even more directly to cause cancer. The sun's UV rays can react chemically with our cells' genetic material, DNA, to produce mutations. The DNA may then send erroneous instructions to dividing skin cells resulting in uncontrolled, cancerous cell growth. One summer of repeated sunburns can increase the danger of cell mutation for each of us by a factor of several thousand.

Just How Strong Is the Link Between UV Radiation and Cancer?

It is believed that roughly 90 percent of all skin cancer is caused by ultraviolet radiation. If you find this hard to believe, consider these facts:

- Eighty percent of the most common types of skin cancer occur on the head, face, neck, arms, and legs, the areas of our bodies most often exposed to the sun.

- Two to three years after a period of intense sunspot activity, which substantially increases our dose of ultraviolet radiation, incidences of skin cancer register a sharp increase throughout the world.

- The incidences of skin cancer are highest in sunny latitudes, especially among fair-skinned populations who have less protective pigments in their skin. Caucasians living in the southern and western states have the highest rate of skin cancer in the United States. One study by the National Cancer Institute revealed that in the Dallas-Fort Worth area, which is situated at latitude 32.8 degrees, more than two-and-one-half times as many nonfatal cancers developed among white males than in the Minneapolis-St. Paul area, which is located at latitude 44.9 degrees. Australia and South Africa are plagued by the highest skin cancer death rates in the world.

- Excessive exposure to the sun caused more than 300,000 cases of skin cancer in 1979 in the United States, according to the National Cancer Institute. This is double the incidence of skin cancer just two decades ago. This rate is attributed largely to the increased popularity of sunbathing and possibly the adoption of more revealing clothes. The effects have been particularly telling on women, who in recent years have been experiencing an increased incidence of skin cancer of the legs.

Can Skin Cancer Be Fatal?

Yes, melanomas, one of the three types of skin cancer caused by ultraviolet radiation, can be fatal. Melanomas happen to be a very aggressive form of cancer, and also unusually resistant to therapy. Scientists have only recently drawn a link between ultraviolet radiation and melanomas, which first appear as molelike growths containing dark pigment. In time, the growths gradually increase in size and bleed easily. They may also ulcerate, and if left untreated they can be fatal within three months. Because melanomas are terminal about 50 percent of the time, they should be treated as soon as they appear.

Fortunately, of the hundreds of thousands of Americans who develop skin cancer each year only about 7500 die from it. This is because most forms of the disease are either basal or squamous types of skin cancer, which are named for the cells from which they develop. Basal cell cancer appears as pale, waxy, pearly nodules; squamous, as red, scaly, sharply outlined patches. Both types can ulcerate and develop crusts. Our basal cells lie at the lowest part of our skin layer and are responsible for more cancers than squamous cells, which compose most of our skin.

Like all cancer, skin cancer manifests itself by uncontrolled growth. When it spreads to other body tissues or organs the cancer is said to metastasize. If not successfully treated, all metastasizing cancers prove fatal. Squamous cell cancers metastasize often, basal cell cancers rarely. Though most skin cancers are not fatal, if a lesion appears, bleeds easily, and refuses to heal, you should immediately consult a dermatologist. It is also important to realize that some ultraviolet-related skin cancers can appear on areas of your skin that have never been directly exposed to the sun, and that not all sun damage is confined to our skin; our eyes in particular can be harmed.

How Does Sunlight Damage Our Eyes?

Both temporarily and permanently. Ultraviolet light acts on the corneas and lenses of the eyes to produce an unpleasant fuzzy bluish light. If our eyes are exposed to too much of this light, the cells in the outermost lining of the eyeball die in only a few hours. The dead cells become opaque and you experience what is commonly termed snow blindness. As the dead cells begin to shed, the eyelid is less well lubricated, and all eyelid movements become extremely painful. Within twenty-four hours you can become photophobic (literally, hating light), since even the slightest amount of light produces small painful movements of your eyelids. Recovery from snow blindness is fairly rapid. A new layer of cells appears within a few days, and within a week the eyes basically return to normal.

We can easily get damaging doses of ultraviolet radiation when snow covers the ground, since snow reflects 85 percent of UV radiation. By comparison, soil and grass reflect only 2.5 percent of ultraviolet light, and beach sand reflects 17 percent. If you're skiing at high mountain altitudes, where ultraviolet radiation is stronger than at sea level, you can become snow blind in three to four hours. This is why Eskimos have developed wooden slit goggles for their eyes. Arc welders and glass blowers working with quartz, which transmits both UVA and UVB radiation, must wear special absorbing glasses to protect their eyes from artificial ultraviolet light. Whether the cause is natural or artificial UV radiation, snow blindness can lead to cataracts, permanent opaque patches on the lens of the eye that must be surgically removed.

Whatever Happened to the Notion that Sunlight Is Beneficial?

It's still valid. But it is important to realize that most benefits derived from ultraviolet light can be gained from normal everyday exposure to the sun, like a short walk or jog, regardless of the season.

Most of us are familiar with the psychological boost sunlight provides, but UVA and UVB rays can also help relieve such conditions as psoriasis, acne, asthma, and aching joints. In addition, ultraviolet radiation is an effective treatment against lupus vulgaris, better known as tuberculosis of the skin in which ulcers develop on the face and neck. As for the notion that UV radiation can cure bone and pulmonary tuberculosis, doctors remain unconvinced.

Artificially produced ultraviolet light also can be beneficial. Experiments in Sweden and Russia have demonstrated that school children and factory workers perform better when exposed to artificial sources of ultraviolet radiation: their stamina is improved as well as their overall immunological responses, particularly their defense against respiratory infections. Just as impressive, though less well documented, are claims that periodic exposure to ultraviolet radiation can lower blood pressure and cholesterol levels, increase blood hemoglobin, improve circulation, rejuvenate nerve endings in the skin, and aid breathing in asthmatics, and in general, increase physical fitness. Most experiments that have demonstrated these results have been conducted on people in northern climates who receive very little sunlight during long winters.

Ultraviolet radiation also enables our bodies to manufacture vitamin D. But we don't have to get much sunlight to produce the vitamin. Even in winter we produce sufficient vitamin D by being outdoors for twenty or thirty minutes with only our faces exposed. Should we have to stay home because of illness or some other reason, we can obtain all the vitamin D we need from foods like eggs, milk, and seafood.

How Does UV Radiation Produce Vitamin D?

Vitamin D exists in several forms. The kind produced by the action of ultraviolet light on the skin is known as vitamin D3, or cholecalciferol. It is manufactured through a photochemical reaction involving UVB radiation. When UVB rays fall on our skin they cause large steroid molecules stored there, 7-dehydrocholesterol, to open their chemical bonds. These stretched-out molecules comprise vitamin D3. In plants, the action of UVB radiation produces vitamin D2, or ergocalciferol, which is vital for growth. As for us, once vitamin D3 is manufactured, it helps our bodies assimilate calcium and phosphorus, which are necessary for healthy bones and teeth. It also activates parathyroid hormones, mobilizes other vitamins, and stimulates enzymes which aid our bodies in operating more efficiently. A lack of vitamin D3 leads to irregular bone development and spinal deformities or rickets in children and osteomalacia or soft bones in adults.

Just as too much ultraviolet radiation is harmful, so is too much vitamin D3. A daily dose of more than a few thousand units taken over several weeks stimulates abnormal buildups of calcium and phosphorus. This excess leads to loss of appetite, general weakness, dangerous deposits of minerals in the heart, aorta, and muscles, and ultimately to kidney failure. Very recent evidence indicates that excessive amounts of vitamin D3 may also lead to the rupture of arterial walls and internal bleeding.

Can I Overdose on Vitamin D3 from Sunbathing?

Actually, by excessively tanning your skin you could experience a slight deficiency of vitamin D3. This may sound contradictory but it's not. While the sun's UVB rays are reacting with steroid molecules beneath your skin to synthesize vitamin D3, the UVA rays are involved in an entirely different chemical process that darkens your skin. The darker your skin, the harder it is for the ultraviolet light to penetrate, and the more difficult it is to synthesize vitamin

D3. In fact a dark brown tan reduces the ultraviolet radiation penetrating the cells beneath your skin by about 90 percent! During the summer this is not a problem because the increased hours of sunlight ensure us an adequate supply of ultraviolet rays for even the darkest tans. But if you return to a northern climate from a winter vacation with a deep tan, your darkened skin may prevent you from absorbing sufficient ultraviolet light during the shorter, often dimmer, winter days. Consequently, you might need vitamin D supplements until your tan fades.

Just What Causes a Tan?

The tanning of skin involves a pigment called melanin. About one cell in ten in the epidermis, or upper layer of our skin, is a melanocyte that produces melanin. Contrary to popular belief, the number of melanocytes in the skin of all people regardless of skin color is essentially the same. Differences in our skin color are due to the fact that more melanocytes are naturally active in the skin of blacks and Mediterraneans, for example, than in the less pigmented skin of albinos. Ultraviolet radiation, primarily UVA rays, triggers already active melanocytes to produce more melanin, hence causing our skin to tan.

The reason our faces, necks, and forearms tan more easily than other parts of our bodies is because these areas contain the greatest concentration of melanocytes. As we age, however, the number of melanocytes in our skin, regardless of skin color, decreases by about 11 percent every ten years. Sunburn damage is believed to account for the disappearance of at least some of the melanocytes, while loss of the rest is probably just a natural consequence of aging. The net result is that the older we get the harder it becomes to acquire either an "immediate" or "delayed" tan.

What's the Difference Between an Immediate and a Delayed Tan?

Well, it's a bit technical, but fairly simple to understand. Regardless of our skin color we all respond to ultraviolet radiation in two stages. In the first stage, or immediate tanning, pale-colored unoxidized melanin granules near our skin's surface are changed by solar radiation to their dark brown form. This action is triggered by both the sun's UVA rays, and surprisingly, by visible light which penetrates even deeper than ultraviolet light. If you find this hard to believe just think back to your childhood when you pressed a flashlight against the palm of your hand and saw a red glow through the other side. You can acquire an immediate tan even through ordinary window glass because the glass transmits both visible and UVA light. The effect is a light coloring of the skin within three or four hours, which fades within a day's time. A more lasting tan results from the second stage, called delayed tanning.

In this process new quantities of pale-colored melanin are synthesized from the amino acid tyrosine in our skin's protein. Ultraviolet light activates certain enzymes which in turn help transform tyrosine into the brown colored melanin. A delayed tan begins about ten hours after exposure to the sun and reaches a peak in approximately four to ten days. It then gradually fades as we shed pigmented cells. In fact, every thirty days our entire upper skin level is renewed. Although we lose a tan in a few months, much of the time a slight coloration remains, indicating that once our skin is sunburned a number of cells are permanently altered.

The amount of permanent damage from ultraviolet radiation depends to a large extent on our skin type and how we sunbathe. Large single doses of UV light, whether from the sun or a sunlamp, are unhealthy for every type of skin. Most people can tan safely, but it is a very individual process. Although there are many variations in skin color, in general, dermatologists recognize four skin types and recommend that we seriously consider our skin types when planning to spend time in the sun.

How Do I Know What Skin Type I Am?

If you are extremely fair-skinned and burn in just ten or fifteen minutes in the summer sun, and if you have never been able to achieve a golden-brown color, you're type I. People in this group have so little melanin in their skin that they run the highest risk of premature skin aging and developing cancerous lesions. If you are of this skin type you should limit your exposure to the sun and always use a sunscreen. It is not surprising that Ireland, with its fair-skinned population, ranks third in terms of skin cancer deaths, even though it receives half the ultraviolet radiation of the two countries that lead in deaths of this type, Australia and South Africa.

People who belong to the skin type II group are somewhat safer. If you have this skin type you usually burn and only rarely tan. But with the proper use of sunscreens you can gradually acquire a faint tan. Remember though, you, too, are highly susceptible to the sun's UVB rays and should never — even with sunscreens — sunbathe during the hours from 10:00 A.M. to 2:00 P.M. when UVB radiation is strongest.

If you're skin type III, you will usually tan and only infrequently burn. But while you can withstand the sun's harsher UVB radiation for longer periods of time, use of a sunscreen will greatly diminish your risk of developing dry, wrinkled skin and cancerous lesions.

If you're swarthy-complexioned like many Mediterraneans, or olive-skinned like many Hawaiians and Puerto Ricans, or black, you're skin type is IV. This means that your skin contains sufficient melanin to offer substantial protection from both kinds of ultraviolet radiation. People with this skin type almost never burn, and easily achieve a rich, dark tan, even after one day's exposure in the sun. But while darker skin offers more natural protection against UV radiation, it still is not immune to serious damage. This is true even on those gray, cloudy summer days when we can't feel the sun on our skin.

How Is It Possible to Get a Suntan on a Cloudy Day?

A cloudy summer day appears dark since not much visible light penetrates the atmosphere, and it may feel cool because the sun's heating infrared radiation is being largely absorbed by clouds. But much of the sun's ultraviolet radiation seeps through the clouds. In fact, up to 80 percent of the sun's UV light can penetrate moderate haze, light clouds, or fog, and about 50 percent can penetrate heavy cloud cover. The reason is that ultraviolet radiation is more energetic than visible light or infrared radiation, so it is not as easily stopped by water molecules in the air.

Since ultraviolet rays are easily reflected upward from the ground, sand, or water, a big hat or beach umbrella is not sure protection. The sun's UV rays can also reach a swimmer underwater and they readily pass through a wet T-shirt and other damp light clothing. Don't be deceived by a cool breeze or low temperature in summer either. You can get just as burned on a sunny June day whether the temperature is 50 or 60 degrees, because the amount of ambient ultraviolet light has nothing to do with air temperature. In fact, because we miss the warning of feeling too hot, we can get a worse burn on cooler summer days. As a general rule in summer, when outdoors for prolonged periods always use a sunscreen.

Are Sunscreens Really That Effective in Blocking UV Radiation?

Some are entirely effective, while others only partially do the job. *Physically active* sunscreens, also called sunblocs, prevent all the potentially carcinogenic ultraviolet radiation — as well as the tanning UVA rays — from reaching our skin. They accomplish this usually by coating the skin with an opaque white substance that permits no ultraviolet radiation to pass through. The two most effective physically active sunscreens are ointments of zinc oxide and titanium oxide.

If you were to use a physically active sunscreen constantly in the sun, you would never tan at all. But by applying one intermittently, you can precisely control your tanning. Unfortunately, physically active sunscreens are messy and unsightly to many people and are therefore usually only practical for small areas of susceptible skin such as our lips and nose.

Chemically active sunscreens can be highly effective without the mess. These sunscreens contain chemicals that slow down or prevent the tanning process by blocking out the sun's burning UVB rays while letting through some or all of the tanning UVA rays. The amount of penetrating rays depends on the particular sunscreen's Sun Protection Factor (SPF), which can range from one, for a very weak screen, to more than fifteen for screens that offer maximum protection. The Federal Drug Administration is attempting to require manufacturers to clearly display SPF factors and tanning recommendations on their products. Here are the FDA guidelines:

• *Ultra* sun protection products, with SPF values of fifteen or greater, should be used for skin type I. These products offer protection from sunburn and permit no suntanning.

• *Maximal* sun protection products, with SPF values of eight to fifteen, should be used for skin type II. They offer maximal protection from sunburn and permit little or no real suntanning.

• *Extra* sun protection products, with SPF values of four to under six, are designed for people with skin type III. They offer extra protection from sunburn and permit limited tanning.

• *Moderate* or *minimal* sun protection products, with SPF values of two to under four, are recommended for people with skin type IV. They offer the least protection from ultraviolet radiation and permit strong suntanning.

Most of these sunscreens use para-aminobenzoic acid as the primary sunscreening agent. It is commonly referred to as PABA, and under varying conditions of moisture and sunlight, and on people of different complexions, a 5 percent solution of PABA in a 50 to 60 percent solution of alcohol has proven consistently superior to dozens of other sunscreens in clinical tests. Unfortunately, many people don't achieve maximum protection from sunscreens because they apply them incorrectly.

How Should Sunscreens Be Applied?

To achieve maximum protection against ultraviolet radiation you should apply PABA solutions to your skin at least one half hour before actually sunbathing. This permits PABA to interact with your skin's oils and enhances its protective properties. In clinical tests, people who applied PABA forty-five minutes before sunbathing found they could take over twice the sunlight before burning. They could also swim farther without it washing off and they required fewer applications.

Although PABA solutions are considered to be the most effective all-round sunscreens, in a few individuals PABA can produce a red irritation of the skin. Fortunately, for these few who are allergic there are other sunscreening agents containing oxybenzone, dioxybenzone, or sulisobenzone. These work well without the irritating side effects.

Within the next few years even more effective sunscreens should be appearing on the market. They are mixtures of physically and chemically active sunscreens that promise to significantly reduce the number of cases of skin cancer from ultraviolet radiation.

I've Read There's a Connection Between the Increase in Skin Cancer and Depletion of the Ozone Layer. Is This True?

There is evidence that the protective ozone layer in the stratosphere is being destroyed by the use of chlorofluorocarbons, and the potential threat of increased incidences of skin cancer is real. Chlorofluorocarbons(CFCs) are chemicals used widely as propellants in many spray cans, as refrigerants, and in foam products. Over the years these chemicals have been gradually percolating into the earth's upper atmosphere where the sun's ultraviolet radiation breaks them down into chlorine atoms. Chlorine has a tremendous appetite for ozone and chews it up in a destructive fashion. No one knows for sure if the amount of ozone destroyed so far has resulted in a higher incidence of skin cancer. We do know, however, that if depletion of the ozone layer continues, an increase in skin cancer is very likely. In fact, many scientists and environmentalists claim that the depletion has already increased levels of ultraviolet radiation reaching the earth's surface, and that the escalation of both fatal and nonfatal skin cancer has been a highly probable consequence.

Scientists are attempting to determine the short-term and long-term effects of a depleted ozone layer on life. By some calculations, for every 1 percent reduction in the concentration of ozone, there is a 2 percent increase in ultraviolet radiation that reaches the earth. A 10 percent reduction in the ozone concentration would, theoretically, increase our exposure to UV radiation enough to cause a 20 to 40 percent rise in the number of skin cancers — anywhere between six to twelve million cases each year in the United States alone.

In December 1979, the National Academy of Sciences warned that if CFC pollution of the stratosphere continues at recent rates, the ozone layer will be reduced 16.5 percent by late in the next century; this is twice the previous estimate. And according to the National Research Council, the ozone threat could be increased if by the end of the century there are fleets of commercial supersonic aircraft flying through the stratosphere. Supersonic planes deposit their chemical exhaust directly into the ozone layer, which could do more damage than present CFC levels. Skin cancer won't be the only harmful result of this imbalance, however; animal and plant life would be threatened as well.

In What Way?

Life on earth has evolved in a delicate balance with UVB radiation levels, and over the years plants and animal cells have adapted to this radiation, learning to repair the slight damage caused by it. An increase in UVB rays would likely overwhelm the biological repair mechanisms of plants and animals that have evolved over centuries, and this imbalance could have a devastating effect on forests, crops, and orchards, and could greatly increase the cases of skin cancer. Many lower forms of animal life would become extinct. Scientists at the National Marine Fisheries Laboratory in Manchester, Washington, have discovered that current stratospheric ozone levels may already be dangerously low, causing damage to certain forms of marine life, such as shrimp and crabs, by allowing more of the sun's UVB rays to penetrate the oceans. Shrimp and crab larvae are already "living close to the thin edge of survival," the scientists reported in 1979, and "if ultraviolet increases, we're going to see the populations of some species die out."

But the largest threat from a substantial decrease in the ozone shield is from UCV radiation, which has not showered the earth for billions of years. If the reduction in ozone were sufficient to permit even small, steady amounts of UVC radiation to reach earth, the effect would be to turn back the geological clock to an era when life as we know it today could not survive.

Then Why Are CFCs Still Being Manufactured and Sold?

There is a worldwide effort underway to cut back on the use of CFCs. This effort has encountered many critics who claim that a decade of scientific research has failed to conclusively prove that CFCs irreparably damage the earth's stratospheric ozone. In fact, according to some calculations recent CFC levels have stabilized since 1974, and other data suggest that the atmosphere can cope comfortably with present levels of CFC pollution.

One of the problems in determining just what's happening to the ozone layer is the fact that it fluctuates naturally. In 1979 scientists at the NASA Langley Research Center in California discovered that the ups and downs in ozone concentration may depend to a large extent on sudden increases in the sun's ultraviolet radiation linked with the 11-year sunspot cycle and unexpected solar storms. An increase in solar ultraviolet radiation, they say, initially converts more stratospheric oxygen to ozone. Yet at the same time, the radiation increases temperatures in the stratosphere, causing ozone molecules to collide with each other at a greater rate than they otherwise would, changing them back into oxygen molecules. The net effect of this circular process is a natural fluctuation in the density of the ozone blanket. Many scientists contend that at the present state of technology it is extremely difficult to measure changes of 5 to 10 percent in the ozone concentrations because of such natural monthly and yearly fluctuations.

Is Anything Being Done to Obtain More Accurate Measurements of Ozone Levels?

There is a new global experiment in which four ozone-monitoring stations have been set up, in Ireland, Barbados, Samoa, and Tasmania. The results of the experiment are expected sometime during the early 1980s. Meanwhile, this country is continuing its 1976 ban on the nonessential use of aerosol sprays, and Sweden is abiding by similar action initiated in 1979.

But it will take more than two countries to cut back on worldwide CFC pollution. Unfortunately, other nations are less willing to impose bans and will probably do so only if it can be conclusively proven that the ozone layer is being seriously depleted. Nonetheless, pressure is mounting in various countries to legislate bans. Many scientists feel that an international ban on CFCs should be imposed until more is learned about the connection among CFCs, ozone, and ultraviolet radiation. They claim that even if release of CFCs into the air were to stop entirely today, ozone depletion would continue for several decades as a result of present levels of the chemicals in the atmosphere.

III The Blanket of Electrical Smog

Microwaves, Radiowaves, and Radar

What Is Electrical Pollution?

*E*lectrical pollution is nonionizing radiation emitted by electrical devices — microwave ovens, radio and television transmitters, and high power lines, to mention just a few. Electrical radiations can be divided into three categories depending on their frequencies. The most energetic kind, microwaves, which are used in microwave ovens and radar, have frequencies ranging from one trillion to one billion cycles a second. They oscillate quickly. Somewhat slower and less energetic are conventional radiowaves that carry television, FM and AM radio, and CB broadcast signals, with frequencies from one billion to roughly one thousand cycles a second. The comparatively sluggish and least energetic radiation is the so-called Extremely Low Frequency (ELF) waves that emanate from high power lines and certain industrial devices. ELF waves possess frequencies below 1000 cycles a second, in the same range as the brainwave patterns of all animals, from earthworms to humans. Since the dawn of the electronics age at the end of World War II, these electrical radiations have been bombarding us in such increasing amounts that it is not an exaggeration to say that we are living in a sea of electrical pollution.

How Extensive Is the Pollution?

A glance at the number of some of the more common electrical devices used in this country will give you an idea of the extent of this new kind of "smog." At the beginning of 1980, in the United States alone, there were 36,000,000 electromagnetic devices in industry for such diverse purposes as curing plywood, sealing the edges of plastic notebooks, and drying potato chips; 30,000,000 CB radios; 9,000,000 broadcasting transmitters; 10,000,000 microwave ovens; 300,000 broadcasting towers; 50,000 radar dishes; 60,000 miles of overhead high power lines; plus thousands of such devices as walkie-talkies, garage door openers, burglar alarms, emergency highway call boxes, medical diathermy machines, surgical cauterizing knives, and the list goes on and on.

As pointed out earlier, if this radiation were visible to the human eye, you could see, for example, the eerie glow surrounding a department store exit that is produced by electronic frisking devices that check for shoplifters. You could watch shafts of "light" relaying telegrams from the roof of one skyscraper to another, or the beacons of "light" transmitted from airports to guide planes in for landings. The amount of electromagnetic pollution from all these electrical sources has skyrocketed to the point that it has actually changed the way the earth looks from space. If in 1900 you could have viewed the earth from space through a telescope sensitive to these nonionizing radiations, you would have seen it as a black sphere. Today it glows like a white sun due to the blanket of electrical smog.

How Is This Electrical Pollution Affecting Us?

No conscientious scientist disputes that these nonionizing radiations are potentially harmful in high doses. For instance, in animal experiments high doses of microwaves have induced cataracts in adults and genetic mutations in fetuses. The damaging cooking effect of microwaves from ovens on the lens of the eye has been proven to be cumulative. This indicates that single exposures to microwave radiation that are not in themselves harmful may become hazardous if repeated often enough. Two questions, however, need to be answered: Can the animal work be validly extrapolated to humans, and is there a level of nonionizing radiation below which there are no significant biological effects? In other words, as with x rays, is there a safe dose?

Until recently most scientists concurred that the levels of radiation emitted by common electrical devices, as well as high power lines, were so low they had no effect on human life. They based their position largely on animal tests conducted in the 1950s by Dr. Herman Schwan of the University of Pennsylvania. From his experiments, Schwan concluded that microwaves damage tissue only by *heating* it, unlike x rays, cosmic rays, and other ionizing radiations that damage cells and chromosomes by ripping electrons off atoms. Schwan demonstrated that in order to produce harmful burns an animal had to receive relatively large doses of microwaves. Such evidence prompted the government in 1966 to adopt a safety standard for microwave and radar exposure of 10,000 microwatts per square centimeter; this is simply a measure of nonionizing radiation exposure falling on a square centimeter of skin or tissue.

At the time, this government limit was one-tenth of the lowest dose known to produce burns in animals. But now there is good reason to believe that this 10,000 figure, which is still this country's safety standard, may be as much as 500 times too high, and the pendulum of scientific opinion has begun to swing in favor of stricter limits.

Why?

Researchers are discovering birth defects, cataracts, cancer, and even behavioral changes in animals exposed to doses of microwaves and radiowaves well within the government's safety limit. Also, there is growing evidence that people might be similarly affected. In light of this later research even Dr. Schwan has cautioned that the safety standard of 10,000 microwatts "badly needs refinements," for several reasons.

First, the standard was formulated without regard to frequency, and there is strong evidence today that lower-frequency radiation penetrates much deeper into our bodies, heating them more intensely than higher frequencies. This means that while a radar maintenance man and a microwave-oven repair man may receive the same radiation doses, the former could suffer more internal damage, since radar has a lower frequency than microwaves.

Second, the current standard was based on simple radiation field patterns, and it becomes practically meaningless for the complex patterns that exist close to radar transmitters and microwave ovens. Furthermore, the main assumption, based in part on Schwan's experiments, that nonionizing radiations like microwaves, radiowaves, and ELF waves damage our cells by heating them, now appears to be wrong. There is growing evidence that such radiations can also cause us harm by rupturing molecular bonds through processes scientists still do not understand entirely.

In fact, this nonthermal damage may actually be more harmful than the thermal damage inflicted on our cells, and many scientists feel that nonionizing radiations might ultimately prove to be a bigger and more vexing problem than ionizing radiations, simply because we are exposed to so many more of them.

What Are Some of the Nonthermal Effects of Electrical Radiation?

Low levels of nonionizing radiation — which are far too weak to cook cells — apparently can affect our central nervous system and trigger our body's stress-response mechanism. Chronic stress, in turn, can lead to heart attacks, strokes, and cancer.

This has been determined from animal experiments, primarily with rats. However, there is convincing evidence as well that people subjected to low levels of nonionizing radiation can suffer similar effects. In one study, Dr. Karel Marha, director of the Department of High Frequency of the Institute of Industrial Hygiene and Occupational Diseases in Prague, studied workers whose occupations expose them to electrical pollution and found that they experienced the same symptoms as people living under conditions of high stress. The workers complained of frequent headaches, exhaustion, dizziness, and sleeplessness at night; they were easily irritated and abruptly changed moods, and became depressed and often paranoid. They were also troubled by muscle and heart pains and experienced irregular heartbeat and breathlessness.

Were These Symptoms Caused by Very High Doses?

Surprisingly not. They were induced by low intensity radiation doses one hundred times less than the U.S. safety standard of 10,000 microwatts. That such low doses could be deleterious was discovered by Soviet scientists decades ago.

In the 1950s Soviet researchers began studying workers — primarily military radar personnel — exposed to prolonged or repeated doses of microwaves or radiowaves, and found that they suffered from headaches, eye pain, malaise, irritability, depression, loss of hair and appetite, and partial loss of memory. The scientists set out to discover the underlying cause of these symptoms and in subsequent experiments demonstrated that people routinely exposed to low levels of nonionizing radiation registered elevated white blood cell counts, a condition symptomatic of an acute infection. The irradiated workers also suffered from a general weakening of their immune systems, which heightened their vulnerability to viruses and infections. Many men became sterile. A few Soviet studies even documented an increased incidence of birth defects among children of these workers. Soviet authorities found the results of these tests so compelling that they adopted a nonionizing radiation safety standard for their citizens of ten microwatts — 1000 times less than the accepted limit in this country — and other East European countries followed their lead.

Are the Soviet Studies Reliable?

For more than a decade the Soviet experiments were dismissed as unreliable by much of the scientific community in this country. American researchers weren't obtaining the same results from their experiments, because the majority of U.S. scientists persisted in their belief that nonionizing radiation had only thermal effects on our cells — and then only at relatively high doses. They knew that nonionizing radiation was not energetic enough to directly fracture chromosomes and genes like ionizing x rays and gamma rays — and certainly no form of radiation was thought to directly modify a person's disposition or behavior. But a small group of American scientists did suspect that the Soviet results might be valid, so they attempted to duplicate the results with their own investigations.

Did the American Scientists Succeed?

Yes. As early as 1963 Dr. Milton Zaret, an ophthalmologist at the New York University-Bellevue Medical Center, suspected a link between microwaves and a particular kind of cataract that forms on the back surface of the eye lens. Organs like our eyes — and to a lesser degree our gall bladder, urinary bladder, the lumen of the gastrointestinal tract, and male testes — are particularly susceptible to microwave damage because they lack a sufficient supply of blood to carry away the damaging heat of microwaves and are therefore easily cooked.

Zaret became suspicious of eye damage after discovering that numerous electronics and communications workers had developed cataracts, and his investigation of these and other cases convinced him that microwaves were the culprits. At the time it had already been proven that microwaves could produce cataracts in the eyes of animals. And they were known to occur in the eyes of workers exposed to intense infrared, or heat, radiation which dehydrates the lens' protein much the way a hot frying pan turns an egg's albumen white. Yet, where typical infrared or heat cataracts form on the front of the lens, Zaret found that microwaves caused cataracts on the posterior surface of the lens.

How Did Zaret Explain the Eye Damage?

He concluded that the cataracts formed for two reasons. First, because microwave radiation penetrates deeply, heating the entire eye not merely the front surface, and second, because the iris, with its greater blood circulation, keeps the front portion of the eye cooler than the rear portion. Zaret discovered that such microwave damage manifested itself early on as a roughening and thickening of the posterior eye surface and that the victim might experience misty vision for years before the actual cataract formed.

By the early 1970s Zaret was sufficiently confident of his findings that he warned the Federal Aviation Administration that the damage of impaired vision due to microwave exposure was real enough to create a safety problem for air traffic controllers and pilots routinely exposed to microwave and radio beams. While scientists are still debating the level of microwaves that can spawn cataracts, the Department of Labor has recently awarded several former air traffic controllers compensation for work-related cataracts, thus acknowledging microwaves' potential for harm.

Is There Any Evidence of Microwave Damage Besides Zaret's?

About the same time Milton Zaret was studying the effects of microwaves on the eye, Dr. Allan H. Frey, a biophysicist at Randomline, a research firm in Huntingdon Valley, Pennsylvania, uncovered evidence supporting the Soviet claim that nonionizing radiation could modify behavior.

By chance, Frey had met a technician from General Electric's radar test facility in Syracuse, who told him that he could "hear" radar. This was considered impossible, but Frey investigated the claim anyway and eventually determined that certain people could indeed "hear" or perceive microwave pulses in the 300 to 3000 cycle per second frequency range used for radar, television, shortwave radio, microwave ovens, and microwave communications towers. He found that people sensitive to the sound variously perceived it as a buzz, tick, or hiss, and that it could be "heard" several hundred feet from an antenna. What's more he discovered that the sounds could be perceived at levels far below the U.S. safety standard of 10,000 microwatts. And, most amazingly, the "sound" could be perceived even by deaf people. So obviously people couldn't hear the waves in the true sense of the word.

Frey soon demonstrated through human and animal studies that the sound of radar could not be heard mechanically by the ear as a result of vibrations of the teeth, eardrums, skin, bone, or muscle. He concluded that some people are able to hear microwaves because the waves directly excite the cells of the auditory portion of their brains. Experiments, in which he attached electrodes to the skulls of cats, bore out this theory. Frey learned that the brain — including the hypothalamic area, where emotions are believed to originate — is highly sensitive to microwave stimulation, and therefore it is quite possible that microwaves could alter our behavior. What was especially alarming was that Frey discovered the brain is capable of reacting to intensities nearly 350 times less than the U.S. safety standard. In addition, Frey and other scientists were accumulating evidence that nonionizing radiation might also disrupt our important blood-brain barrier.

What's the Blood-Brain Barrier?

The blood-brain barrier is a vital system of capillaries surrounding the brain that maintains it in a stable environment by allowing certain proteins and sugars in the blood stream to enter the brain while blocking other substances out. In his experiments, Frey and his colleagues injected rats with a fluorescent dye that binds to proteins in the blood stream, which do not normally penetrate the blood-brain barrier. Once the rats were exposed to extremely weak microwaves, however, the proteins entered the rats' brains, clearly indicating that the protective barrier had been damaged.

Scientists are trying to determine if microwaves can similarly affect people. If so, this permeability of our brain's screening membranes might allow our blood to transmit the many additives and preservatives in our food, as well as bacteria and viruses to the brain, disrupting its normal functioning.

Wasn't There Some Alarm a Few Years Ago About American Officials in Moscow Being Affected by Microwaves?

Yes, there was. In 1962 the United States State Department discovered that the American Embassy in Moscow was being irradiated by a microwave beam. This irradiation continued through the 1970s, with the beam reaching a maximum intensity of eighteen microwatts, 500 times less than what the State Department considered harmful, but nearly twice the intensity the Soviet government permits for its own citizens. The beam was presumed to be part of an elaborate jamming and bugging scheme, although some pentagon officials speculated it was an attempt to alter the behavior of American personnel.

Whatever the reason, the State Department decided not to inform the embassy employees that the building was being irradiated. In 1966, concerned about the possible health implications of the beam, the State Department launched a secret study of the effects of low-level microwave irradiation, Project Pandora. By the mid 1970s, government officials had concluded that the continual irradiation of embassy personnel had no adverse biological consequences, and the project was abandoned. Shortly thereafter the employees in the embassy were officially notified of the radiation.

By this time, however, Ambassador Walter Stoessel, whose office caught the center of the beam, was reported suffering from a blood disorder, nausea, and bleeding in the eyes. Stoessel resigned the post in September, 1976, for health problems.

Did the State Department Then Recognize the Danger of the Soviets' Microwave Beam?

No. While the State Department installed metal screens on the windows of the embassy and declared the post "unhealthy . . . based upon the Department's evaluation of reported environmental conditions regarding sanitation and disease, medical and hospital facilities, and climate," it never directly mentioned radiation. Indeed, the department went out of its way to reassure embassy employees that no medical problems had been identified with low levels of microwaves. Even when the American Embassy Post Medical Officer detected an abnormally high elevation in employees' white-blood-cell counts — a condition called lymphocytosis, characteristic of the body's response to acute infection — the staff was advised that this condition was probably due to an unknown virus in the Russian atmosphere, and definitely was not the result of microwave irradiation. What the employees were not told was that more than a decade earlier, Soviet scientists had discovered that lymphocytosis could be induced by microwaves.

Today, Stoessel is living in West Germany, and, according to reports, he is suffering from lymphoma, a form of leukemia. If this is true, Stoessel's condition assumes added significance in light of the fact that two of his predecessors, Charles Bohlen and Llewellyn Thompson, died of cancer.

Has a Link Been Established Between Microwave Radiation and Cancer?

To date, no firm link exists. Many scientists regard the likelihood of microwaves causing cancer as nil. But others point out that though microwaves are nonionizing, and thus theoretically noncarcinogenic, nonionizing ultraviolet light does induce skin cancer. Despite such differing views, there is disturbing circumstantial evidence suggesting a connection. The real problem is that any pertinent data must come from retrospective studies of people exposed to nonionizing microwaves and radar, since, for obvious health and ethical reasons, people cannot be exposed experimentally. As such, the data are difficult to evaluate, but here is a portion of the circumstantial evidence.

• In 1978, the chief epidemiologist for New Jersey discovered that the cancer rate in the town of Rutherford was far above normal. Five children who attended the Pierrepont Elementary School had developed cancer, and the odds against this happening were estimated at ten million to one. Coincidentally, about 6400 sources of microwaves — including broadcasting transmitters and radar antennas — are located within a fifteen-mile radius of the town according to the Federal Communications Commission. Furthermore, the school is situated on top of a hill, which makes it an easy target for microwaves. When the citizens of Rutherford learned of the problem, they insisted on microwave measurements, and the National Bureau of Standards performed the job. The bureau found unusually high field strengths throughout the town and currently other government agencies are planning to examine the people of Rutherford.

• One case of skin cancer and two cases of leukemia were found among seventeen men between the ages of thirty and fifty who had been conducting radar tests at United States missile sites between 1968 and 1972.

- Of eight civilian employees at a naval air station in Rhode Island, three recently developed cancer. All three men were responsible for repairing radar equipment. One man died of lung cancer in 1970 and another died of cancer of the pancreas, lung, and liver in 1973. The third man has also developed cancer of the pancreas, lung, and liver. What is curious and lends weight to the arguments that these cancers were occupationally induced is the fact that although pancreatic cancers are very rare in people under the age of forty, both victims were in their thirties.

- A survey in the late 1970s disclosed that sixteen American women serving in the Moscow embassy during the time it was irradiated developed breast cancer.

Speaking of the Dangers to Women, Wouldn't a Developing Fetus Be Particularly Vulnerable to Microwaves?

Yes. Although the dangers of low-level nonionizing radiation have not been proven to the satisfaction of the scientific and medical communities, it is reasonable to argue that if any living organism on earth were sensitive to this radiation it would be the developing fetus.

We know that animal cells exposed to varying intensities of microwaves develop chromosome damage, and animal embryos irradiated with radiowaves can be either stillborn or born with gross abnormalities. Claims have also been made that microwaves can similarly affect us. There are reports of a high incidence of mongolism among children of airline pilots and air-traffic controllers, who are exposed to microwaves. And there are claims that children born to families who live on military bases containing radar installations are more likely to experience genetic defects, such as cleft palates, clubfeet, and anomalies of the heart, genital organs, and circulatory and respiratory systems; also, that fetal death rates are abnormally high among families located near radar complexes.

Still, evidence that low-level nonionizing radiations produce such mutagenic effects in humans is only tentative to date and it is a hotly debated issue. Researchers at Johns Hopkins University have demonstrated that men occupationally exposed to radar and microwaves have a significant increase in the number of chromosome abnormalities in their blood, though these abnormalities have not yet been linked to birth defects. At George Washington University scientists have also studied young women returning to the United States who worked in the Moscow embassy. Out of 140 blood samples examined over a four-year period, four of them showed seriously defective chromosomes.

Isn't the Case Against Microwaves Pretty Convincing?

It does sound that way. Yet for every expert who contends that microwaves are harmful in doses even lower than the national safety standard, another expert claims these waves are innocuous even at exposures over the national standard. Because microwaves and radiowaves weaken quickly after they leave their source, some scientists argue that simply stepping back from a police radar gun or a microwave oven, for example, reduces your exposure ten to one hundred fold. Others contend that such precautions aren't necessary since the animal studies with microwaves are not applicable to humans. Still others claim that the reports of increased mongolism, congenital abnormalities, and fetal death rates in children of pilots and radar technicians don't hold up under close examination. Only recently a study by the National Academy of Sciences found that naval radar operators died no younger than their peers in various jobs, and the Environmental Protection Agency points out that 98 percent of the American population is exposed to less than one microwatt of microwave radiation at any one time.

Those who cast microwaves in a less benign light are not convinced by these arguments. One of the loudest voices raised against microwaves is that of journalist Paul Brodeur. In his controversial 1977 best seller, *The Zapping of America: Microwaves, Their Deadly Risk, and the Cover-Up,* Brodeur charges that "Microwave radiation can blind you, alter your behavior, cause genetic damage, even kill you. The risks have been hidden from you by the Pentagon, the State Department, and the electronics industry." Brodeur claims that many proposed studies to examine the mutagenic potential of low levels of nonionizing radiation have been either endlessly delayed by red tape or denied government funding, and at times damning evidence has been suppressed.

Is There a Coverup Concerning Microwave Radiation?

Certainly enough money and equipment is at stake to make the coverup theory plausible. Imagine what would happen it it were proved that levels of electrical pollution found in communities situated near microwave and radar installations were sufficient to damage fetuses as well as children and adults.

For starters, the armed forces could be sued by every military and civilian employee who had been exposed to radar and subsequently developed health problems or had an abnormal child. Companies that manufactured and tested microwave equipment — from communication antennas to ovens — would be vulnerable to similar legal action by employees who contracted diseases ranging from cataracts to cancer. In industry, the lucrative microwave oven business would fold overnight.

What's more, an enormous segment of our nation's radar and weapons defense systems operate at radiation levels just below the current safety standard. If new experiments proved that standard unsafe and the standard had to be substantially lowered, the cost of relocating radar and missile sites away from populated areas, and of perhaps even redesigning safer equipment, would run into tens to hundreds of billions of dollars.

Obviously the government, the military, and industry have much at stake, and there are surely individuals within each sector who would be willing to soft-pedal any dangers presented by microwaves or radiowaves, to protect their own interests.

But High Stakes and Conjecture Are Not Proof of a Coverup.

That's true. The government's inaction is more the result of genuine confusion than conspiracy. Scientists are honestly bewildered over the insufficient and often conflicting evidence concerning the biological effects of low levels of nonionizing radiation.

For one thing, the study of the effects of nonionizing radiation — particularly radiowaves and extremely low frequency waves, like the kind we use for electrical power lines — is in its infancy. In addition, the setting of standards has never exactly been on the cutting edge of research. Despite the existence of several federal research programs on microwaves, radiowaves, and ELF, the subject of the hazards of these radiations has only recently begun to attract the scientific attention it deserves.

The government has taken an encouraging first step by budgeting nearly $10 million a year for more than one hundred research projects to investigate the effects of low levels of nonionizing radiation. Also, an increasing number of American researchers have come to realize that the United States safety standard of 10,000 microwatts rests on the unproven assumption that microwaves and radiowaves damage biological tissues only by heating them. In fact, recent evidence that low levels of nonionizing radiation may affect our central nervous system in the same manner as chronic stress, weakening our immune defenses and perhaps also altering our moods and behavior, has begun to swing the pendulum of scientific opinion in favor of those who advocate stricter limits. As a result, there is mounting pressure on federal agencies to reduce the acceptable limit of occupational exposure to nonionizing radiation.

Is the Government Planning to Revise the Current Safety Standard?

The federal government's National Institute for Occupational Safety and Health (NIOSH) is proposing *mandatory* safeguards stricter than the current *voluntary* measures, though the exact limit has yet to be determined. And the American National Standards Institute, a private association of government and industry officials is planning to lower its maximum recommended exposure for workers from the current 10,000 microwatts to 1000 microwatts. However, a growing number of scientists feel that until we are more certain of the health risks posed by microwave exposure the Maximum Permissible Dose to workers should prudently be set at 100 microwatts — and at half this figure for the rest of the population.

Other countries are equally concerned over the possible negative effects of low levels of nonionizing radiation. The Canadian government recently lowered the maximum allowable exposure for Canadian workers from 10,000 microwatts to 5000. And the Soviet government has proposed to reduce its already tight limit of 10 microwatts exposure to the general public to a one to five microwatt dose.

Are Many Americans Exposed to On-the-Job Microwaves and Radiowaves?

More than you might suppose. According to the National Institute for Occupational Safety and Health, twenty-one million of us work in the vicinity of some kind of microwave or radiowave device. Many people, for example, work with industrial sealers which employ molds heated by electric currents in the radio and microwave frequencies to melt plastic into various shapes to produce consumer goods such as shower curtains, tablecloths, and hundreds of other plastic items.

According to one NIOSH study, many workers are being exposed to limits above the current dubious safety standard. Between 1974 and 1978 the institute launched an investigation in which it visited twelve industrial plants and measured the exposure levels near eighty-two plastic sealers. The tests showed that more than 60 percent of the sealers emitted radiation above the recommended safety limit. In fact, some machines exposed their operators to field intensities of twenty-six times the safety standard, actually heating the workers from within.

What is particularly disturbing about this discovery is that most of the operators were women of childbearing age, and no one knows what effect this radiation could have on their offspring. None of the workers was ever aware of the electrical pollution because microwaves and radiowaves easily pass undetected through those layers beneath our skin that contain nerve endings, depositing their heat, and possibly breaking chromosomes.

It is unfortunate that any workers are exposed, because they don't have to be. For instance, many industrial employees could be protected by metal shields which manufacturers already provide for machines sealers in several countries that have stricter mandatory industrial safety standards than the United States.

If You Don't Work with Radar Devices or Cook with a Microwave Oven You're Fairly Safe?

Don't kid yourself. Do you live near a radio or television broadcasting station? Within several miles of a satellite communications antenna or a military radar installation? Do you operate a CB radio? Or do you shop in a department store that has installed an electronic frisking system at entrances and exits? Tens of millions of us could answer yes to these questions, and that means we are being exposed to microwaves.

The Environmental Protection Agency's office of Radiation Programs has discovered just how extensive the pollution is. The EPA has measured electrical radiation levels at more than a hundred locations in Atlanta, Boston, Miami, New York, Philadelphia, Los Angeles, Maryland, and Washington, D.C. The department has found, for example, peak radiation measurements between 50 to 100 microwatts on the fiftieth floor of the Sears Tower in Chicago, on the thirty-eighth floor of the Biscayne Tower in Miami, and on the forty-seventh floor of the Milam Building in Houston, and a level of ten microwatts on the fifty-fourth floor of the Pan Am Building in New York City. And the powerful television transmitters on top of one tower of Manhattan's World Trade Center, which have begun to be turned on, could expose millions of sightseers and workers on the upper floors of the companion tower to as much as 360 microwatts. All of these amounts exceed the Soviet safety standard.

Intensities of several hundred microwatts have been detected under a radar unit on small boats and near police traffic radar. Furthermore, exposures above the U.S. 10,000 microwatt limit have been measured within five inches from hand-held walkie-talkies and within six inches from electrical cauterizing knives used in surgery, which are usually directly aligned with the surgeon's eyes. The two-way radios in taxicabs, ambulances, trucks, and police cars can radiate fields around nearby cars of more than 60,000 microwatts, and, according to a government survey, radar repairmen may work in fields as high as 100,000 microwatts.

Admittedly, many of us experience these high doses only for brief periods of time, and the lower doses measured on the upper floors of skyscrapers are well below what most scientists consider to be potentially dangerous levels. In addition, there is no proof that sporadic bombardment is detrimental to our general health. Nevertheless, such measurements do illustrate how ubiquitous electrical pollution is and why we must be concerned about it, particularly in light of the fact that technological breakthroughs in the next twenty years will be introducing us to many new devices that will further thicken the blanket of electrical smog that already covers us.

What Are Some of These Devices?

There are three major devices that will be widely used in the near future: a new Microwave Landing System due to be installed at every major airport throughout the world; solar-powered satellites that will beam the sun's copious energy to earth in the form of microwaves; and car phones so inexpensive, versatile, and convenient that just about every driver is expected to own one; the phones will operate by relaying calls from car-to-car or home-to-car via radiowaves emitted from scores of transmitters situated throughout a city.

These devices will offer us tremendous benefits, but there is no denying the fact that their existence could more than double our present level of electrical pollution. And this estimate does not include microwave ovens which during the 1980s are expected to become a standard household item. Already we have bought more than ten million ovens for our homes, restaurants, and institutions, and sales are expected to climb at three million a year throughout this decade.

Just How Safe Are Microwave Ovens?

The most immediate health consideration centers on the ability of microwaves to produce cataracts — and the possibility that the effects of repeated low doses are cumulative. Since many of us periodically peer through the window to check on the progress of cooking foods, our eyes are often exposed to microwave radiation. If dose effects prove to be cumulative, this would be a very dangerous practice, especially considering the potentially more serious effects of chromosome damage, and the suspected trait of microwaves to trigger our body's stress mechanism.

All commercial microwave ovens operate at a frequency of 2.45 billion cycles a second. While this is at the lower end of the microwave portion of the electromagnetic spectrum, just slightly above radiowaves, researchers have been able to produce cataracts in rats and monkeys with this frequency wave. Dr. Ezra Berman, of the Environmental Protection Agency's Health Effects Research Laboratories, bombarded pregnant mice with low microwave frequencies and produced a significant number of offspring with ezencephaly, a birth defect commonly known as hernia of the brain. In his experiments, Dr. Berman irradiated the pregnant mice with doses as low as 3400 microwatts. The current maximum allowable radiation leakage from microwave ovens is 5000 microwatts — although newer and properly operating ovens leak significantly lesser amounts. While this is half the U.S. safety standard, it is more than adequate to induce a variety of aberrations in animals. In addition to animal experiments, there is other recent evidence to document the potency of the lower microwave frequencies emitted by microwave ovens.

What Evidence?

In December, 1979, British astronomers found that radiation from microwave ovens could interfere with radiotelescopes used in the scientific study of space. Astronomers at the University of Manchester's renowned Jordell Bank Observatory calculated that leakage within the British government's limit for a single microwave oven increases twelvefold the background interference for a radiotelescope just one-half mile away. In practical terms, this means that a radiotelescope tracking an object in space could be rendered inoperable by a housewife cooking a roast in her microwave oven fifteen miles from the observatory.

Taking into account the anticipated proliferation of microwave ovens in this decade, the Jordell astronomers have cautioned that the ovens represent a "serious danger" to the whole field of radioastronomy. The British Microwave Oven Association, whose job is to police emissions from these appliances, has suggested that the radioastronomers relocate their telescopes away from densely populated areas, to the Scottish highlands for instance. An astronomer may have such a choice, but a housewife who has just put her family's evening meal into her microwave oven can't very well avoid radiation leakage by retreating to the highlands until it is cooked.

Are Oven Manufacturers Trying to Reduce Leakage?

Manufacturers maintain that by adopting an exposure limit that is one-half the U.S. standard, they have already adequately protected the public. Still, they are attempting to reduce leakage as much as possible. A recent survey revealed that the most modern ovens leak no more than an average of 120 microwatts — and this at only a few inches from the door. Since virtually all microwave ovens leak a certain amount of radiation in their lifetime, and since their safety cannot be absolutely guaranteed, manufacturers and scientists alike agree on specific precautions you should take when operating them.

What Are the Precautions for Using a Microwave Oven?

- Strictly follow the manufacturer's instructions. For example, do not use metal cookware because it reflects microwaves. If your oven already leaks, metalware will increase the radiation that enters the kitchen.

- Check to make sure your oven was not damaged in shipping. A dent in the side, for instance, could be sufficient to render the safety seal around the door useless.

- Don't watch foods cooking or loiter in front of the oven for long periods of time.

- Never stick objects through the door grill or around the seal.

- Don't operate an empty oven. When there is no food to absorb the radiation, more of it can leak into the room.

- Don't tamper with or loosen the door interlock.

- Don't use abrasive detergents to clean your oven. Instead, use water and a mild detergent.

- If your model was produced before the tighter 1971 safety standards took effect, turn the oven off before you open the door and stay at least two feet away from the oven when it is on.

- Thirty-six thousand microwave ovens manufactured by General Electric Company between November, 1973, and October, 1975, are suspected of leaking radiation in excess of the 5000 microwatt standard. These were recalled in 1977, and General Electric agreed to repair the ovens free of charge. No oven manufactured in this interval that has not been repaired should be operated.

- Finally, have your oven checked periodically at a qualified service station authorized by the manufacturer or check your local health department for a reputable dealer. There are also instruments available that enable you to perform your own spot check. One, Micromate, is a small hand-held unit that detects dangerous levels of leakage. *Consumer Reports* rated the Micromate to be quite sensitive over a narrow range of leakage. The magazine recommends, however, that it be used only as an interim safety check and not as a replacement for a periodic check by authorized personnel.

Is the Food Cooked in a Microwave Oven Contaminated by the Radiation?

No, this is a misconception. Food cooked in a microwave oven is perfectly safe to eat. It's the cooking process that can be dangerous for anyone in the vicinity of the oven. In a microwave oven, water molecules in food are forced to vibrate rapidly, producing heat which cooks food from the inside out. Naturally the cells in a stalk of broccoli or an eye roast you are cooking are severely damaged: chromosomes rupture, genes fracture, and the hereditary DNA chains split into thousands of pieces. But the damage is only from the heat and is exactly what happens to food in a conventional oven. The only radiation in the food taken from a microwave oven is this heat, which is no more harmful to us than the heat from a light bulb, the sun, or warm water.

ELF Waves, High Power Lines, and the "Weekend Effect"

How Do ELF Waves Compare with Microwaves?

Like microwaves, ELF (extremely low frequency) radiation is nonionizing, less energetic, and therefore can't destroy our cells by ripping electrons off atoms. However, while the two radiations are similar in this respect, they also differ in a very important way. ELF radiation, which is emitted by high power lines, industrial machinery, and many household items, and characterized by frequencies roughly under 1000 cycles a second, cannot cook our body tissues or rupture molecular bonds, as microwaves are suspected of doing. They can, however, affect our body in a much subtler way, particularly those waves in the ½ to 100 cycles per second range.

This is because our body cells and organs, like those of every living organism on earth, have their own electromagnetic fields that lie within the lower ELF range. There is strong evidence suggesting that these natural bodily electromagnetic patterns, such as the vital electric currents of our heart, which has a frequency of sixty cycles per second, or our brain waves, with frequencies from one-half to thirty cycles per second, can be disrupted when ELF waves interact with them.

What Accounts for Our Body's ELF Waves?

It's believed that ELF radiation provided the energy that created life on earth more than three million years ago. Long before technology polluted the environment with electrical smog, the earth's electromagnetic atmosphere was very simple. The low end of the electromagnetic spectrum consisted of a band of radiation with frequencies from one to thirty cycles a second, which was produced by resonance among the earth's surface, the earth's natural magnetic field, and the ionosphere. The only other low frequency radiation of any magnitude was on the 1000 cycle a second range, produced by lightning discharges.

The remainder of this spectrum was empty. Modern-day analysis of the low-frequency radiation that existed when life began on earth indicates that enormous amounts of energy were present in the ELF region, particularly centered around the ten cycle a second frequency. This phenomenon is known as the Schuman resonance after its discoverer and was probably the source of the energy required for the construction of the complex biological molecules that finally resulted in life. It is believed that all life began and has evolved in this simple electromagnetic environment, which explains why all living things operate on their own electromagnetic waves.

Our own electromagnetic fields, as those of other animals and plants, interact with the earth's natural ELF fields. The biorhythms of many organisms are clocked to oscillations of the earth's ELF fields: migrating birds use them to orient themselves; and people placed in electrically shielded cages that block out ELF frequencies experience changes in body temperature as well as disruption of the excretory function of the kidneys and the chemical composition of urine in a matter of days because their body rhythms go askew.

Until the advent of the Electrical Age, the earth's natural electromagnetic environment had remained relatively constant, forming an intricate link with all life. The concern now is what effect the vast increases of manmade ELF radiation from electrical devices will have on this delicate balance.

Do All Electrical Devices Emit ELF Radiation?

Yes. If for the sake of comparison we assign the number one to the strength of the electric field measured twelve inches from an electric blanket, then the strength of the field at the same distance from a plug-in broiler oven would be two, from a refrigerator or iron four, from a toaster five, and from a light bulb ten. The field strengths from these devices, however, are extremely weak and represent no known health hazard. But high power lines are a worrisome source of ELF radiation and there are roughly 60,000 miles of lines throughout the country.

In understanding the potential problem it is important to realize that an electric field does not flow only in the direction of a power line, but also perpendicular to it. You can prove this by holding a fluorescent bulb a few feet under a high-voltage line and watching the bulb light up. ELF fields can extend to thousands of feet and are so strong, in fact, that if you touch a car parked directly under a high power line you can receive a mild shock because the line's radiation is electrically charging the car.

As our use of electrical power continues to expand, new ultra-high-power lines are going to be used that will carry several million volts of electricity. Already the older 500,000 volt lines are being replaced by 765,000 volt ones. The ultra high lines will radiate ELF fields for over a mile in all directions. In addition, several military communications systems are currently under construction that will transmit data via ELF waves. This proliferation of ELF radiation is drawing increasing attention from health specialists, and recent experiments show cause for concern.

What Are the Effects of ELF Radiation?

Although far too few studies have been conducted with manmade ELF radiation to concretely determine its biological effects, there are numerous indications that such radiation can be harmful. For instance, ELF fields have been found to slow the heartbeat of eels and salmon, and homing pigeons exposed to the radiation become disoriented. Bees placed in manmade ELF fields have stopped storing honey and pollen and have even killed themselves, sealing off their hives and suffocating to death.

At the Upstate Medical Center in Syracuse, New York, Dr. Robert Becker, a pioneer in the study of the biological effects of ELF fields, has concentrated his research on the impact of sixty cycle waves radiated from high power lines. He exposed mice for thirty days to ELF strengths emitted from high power lines and discovered they experienced changes in their hormones, body weight, and blood chemistry similar to those found in animals subjected to chronic stress. Becker has since raised three generations of mice in continuous ELF environments. In addition to displaying all the symptoms of chronic stress from weight loss and stunted growth to hypertension and degenerative diseases of the organs, the infant mortality rate in the third generation was 50 percent compared to the normal rate of less than 5 percent.

Is There Any Indication ELF Has the Same Effect on Us?

Becker feels there is no reason to believe that mice are any more sensitive to electromagnetic fields than humans and suspects that sixty cycle fields act directly upon the human brain, without being consciously perceived, triggering our stress response mechanism.

In addition, Soviet studies have indicated that ELF fields near our heads can induce mild anesthesia and alter perception and attentiveness. Dr. Ross Adey of the Brain Research Institute of the University of California at Los Angeles has achieved similar results by exposing cats, monkeys, and people to ELF fields ranging from two to fifteen cycles a second, a maximum of only half the brain's natural spectrum. Adey has found that as a result both animals and people experience slower than normal reaction time to learned tasks. If his conclusions are correct, this means that people who work around industrial equipment generating ELF fields would risk altering their perception, attentiveness, and reaction time, and possibly jeopardize their safety.

Does ELF Radiation Affect the Brain by Distorting Its Own Frequencies?

That's exactly what happens. This is believed to occur through a process called entrainment. External ELF frequencies identical to those located in the brain — primarily from one-half to thirty cycles a second — either ripple in unison with the brain's waves, reinforcing them, or ripple contrary to them, diminishing their energy. Either way our brain's natural patterns are altered. If the ELF waves couple harmoniously with our brainwaves they are believed to actually enhance our perception and sharpen reaction time, while nonsynchronous interference has the opposite effect. But at present it would be a tricky, risky business to attempt to achieve the apparent beneficial effects while avoiding the possible harmful ones.

There may also be a biochemical mechanism involving the brain, as Adey discovered while working with his cats.

What Did the Cat Experiments Reveal?

That the ELF fields affected the distribution of calcium ions in the cats' brains. Apparently the radiation literally shook the ions from their natural locations on cerebral nerve cells. Calcium ions are charged chemicals that play a vital role in transmitting signals throughout the brain. Adey has tentative evidence that frequent exposure to nonionizing ELF radiation also alters an animal's natural sleep patterns.

Other scientists have speculated that manmade ELF fields — as well as freak ELF fields produced by electrical storms — may at times be responsible for car and plane accidents, by altering a driver's or pilot's reaction time. Some people evidently are more sensitive to ELF fields than others. According to a report published in the March 1979 *American Journal of Epidemiology*, in a large residential area of Denver, Colorado, the greatest numbers of children with cancer lived in homes located nearest to power line transformers. These children were continually exposed to ELF fields; however, these findings are not conclusive, and no one can yet describe with certainty the full spectrum of effects from ELF radiation. If the potential for harm is confirmed, though, the most serious threat to us may come from high power lines, particularly on weekends.

On Weekends?

It does sound strange, but it's true. In late 1975 engineers at Stanford University began reporting the existence of puzzling abnormally large electrical disturbances in the earth's ionosphere and magnetosphere that occurred only on weekends. The ionosphere is the electrically charged region of the upper atmosphere that blankets the earth and reflects radio signals for long-distance communications. The magnetosphere is a region beyond the ionosphere in which the earth's magnetic field lines connecting the north and south poles stretch far out into space. Both these spheres protect us from radiation emitted by solar storms and bombardment by cosmic rays.

The disturbances detected by the Stanford engineers were capable of interfering with long-distance radio communications on earth, and, in effect, poking holes in the earth's protective blanket. Since there was no known weekend electromagnetic cycle that occurred in nature, Stanford scientists conjectured that the effect must be manmade, and that it probably was the result of leakage from power lines which store and radiate more power on weekends when closed office buildings and companies require less power.

Have Scientists Proved that the "Weekend Effect" Is Manmade?

By demonstrating that the disturbances in the magnetosphere center at a frequency of fifty cycles a second over Europe and sixty cycles a second over North America — corresponding to the different predominating power-line currents used on the two continents — they have confirmed that the weekend effect is a manmade phenomenon.

Scientists are speculating how the weekend effect might affect life on earth — or even the earth's weather. A Canadian power company is currently measuring the radiation from its transmission lines that leak into the magnetosphere to get a better idea of the interaction between high power line leakage and the earth's magnetosphere. Japanese researchers have proposed sending balloons, rockets, and satellites into the magnetosphere to monitor the effects of ELF pollution.

However, this area of research is very new and we probably will be well into the decade before we know more about the consequences of the weekend effect, as well as other forms of ELF radiation, for us. In the meantime, since we cannot very well stop using ELF waves or microwave and radiowave energy, for power and communications purposes, it might be prudent to refrain from producing any new nonessential sources of electrical pollution, or at least try to keep as many of its sources as possible away from populated areas, until research results enable us to clearly determine the potential hazards, and decide when they are outweighed by the benefits.

IV Radioactive Garbage

The Deadly Pileup:
Where Can We Put It?

What Exactly Is Radioactive Garbage?

It is radioactive waste that comes from the production of nuclear weapons, the generation of nuclear power, and the use of radio-isotopes in science and industry and in medicine to treat disease. Many of these so-called radwastes present a potentially serious problem for us, not only because they are dangerous sources of long-lived radiation, but because they are piling up at a fantastic rate. And believe it or not, we are thirty-five years into the nuclear age and still have not devised a safe, permanent means for storing the deadliest of them.

Until the mid 1970s, most radwaste came from the production of nuclear bombs and warheads and the generation of nuclear powered submarines. As a result, we are stuck today with more than a half-million tons of highly radioactive weapons waste, in addition to over sixty million cubic feet of lower-level weapons debris. South Carolina, Washington, and Idaho, alone, the sites of three major federal radioactive waste dumps, contain more than nine million cubic feet of waste from the reprocessing of radioactive plutonium for the production of nuclear weapons. While weapons waste has been the major contributor to the radwaste build-up to date, waste from nuclear power plants and medical facilities is quickly catching up and will soon surpass it. The waste from nuclear power plants is of particular concern to us.

Is Nuclear-Power-Plant Waste Different from Weapons Waste?

Yes, waste from a nuclear power plant contains more radioactivity than weapons waste, and it is being churned out by the world's more than 230 nuclear power plants at an alarming rate. Every day each one of these plants produces tens of thousands of cubic feet of potentially hazardous waste, and there are another 233 plants under construction. The waste from this country's nearly seventy operating nuclear power plants alone is expected to increase sevenfold in the next ten years, and by some estimates will total nearly 200,000 metric tons. According to the Environmental Protection Agency, by the year 2000, the lower-level radioactive reactor waste alone will total nearly one billion cubic feet, enough to cover a four lane highway from coast to coast with one foot of radioactive debris. And there will be tens of thousands of tons of additional highly radioactive waste.

To make matters worse, the reactor-site pools built to hold the waste are rapidly being filled to capacity. Studies by both industry and government show that at least twenty-eight nuclear power plants in this country could be forced to shut down by 1986 because of lack of storage space for spent uranium fuel. Even if a moratorium on nuclear power generation were declared tomorrow, we already have huge stockpiles of this lethal garbage to dispose of. There can be no denying that the accumulation of radioactive wastes presents us with one of the toughest problems of the nuclear age, as well as a potentially serious health threat because of the deadly radiation they emit.

Does Radwaste Radiation Damage Our Cells the Same Way X Rays Do?

Yes, like x rays the radiation emitted by radwaste is the ionizing kind that rips electrons off atoms, producing highly reactive particles called ions that can cause cancer and birth defects. But radwaste radiation is potentially more dangerous than the radiation we receive from medical x rays because it mainly takes the form of *nuclear particles* and waves rather than just pure wave radiation like x rays.

These particles and waves are emitted by the nuclei of unstable radioactive atoms in the form of alpha and beta particles, neutrons, gamma rays, and various combinations of these. The reason radioactive atoms give off these emissions is that they are continuously trying to shed their radioactivity so they can achieve a stable, nonradioactive state. We find some of these radioactive substances in nature, but by far most of them are manmade.

Are Some of These Nuclear Radiations More Dangerous than Others?

They all can seriously harm us, but certain kinds do more damage than others, depending on how densely they ionize, or destroy, our body tissues. In general, beta particles and gamma rays are as damaging to our cells as x rays, while neutrons inflict ten times as much damage, and alpha particles twenty times as much destruction as x rays.

Actually, alpha particles travel only extremely short distances. Therefore, if they originate from radioactive material outside our body, they can do little damage since they cannot penetrate the horny outer layer of our skin. But alpha particles can cause serious harm if we inhale or swallow radioactive elements that emit them, or if these elements enter a wound or puncture in our skin. This can happen when radioactive waste contaminates our air, soil, and plants. And when these alpha-emitting elements work their way into our body, the damage they do is highly concentrated in the tissue immediately surrounding the radioactive material and it can easily result in cancer.

Like alpha particles, beta particles from external sources do not penetrate far into our body and they are much more dangerous if we ingest the elements that emit them. Although they often do possess energies sufficient to drive through our skin to depths of more than half an inch. In terms of the ability of particle radiations to harm us, then, beta particles lie between alpha rays, which do highly concentrated damage along very short paths, and neutrons which penetrate deeper into our body with damage thinly distributed along the line of penetration.

So the Kind and Level of These Nuclear Radiations in Waste Determine How Harmful It Is?

Right. Generally wastes that generate low levels of heat and radioactivity are called low-level wastes. These are mainly produced by hospitals, research laboratories, and by nuclear power plants in the form of contaminated machinery and materials in the plant. On the other hand, wastes that emit intense heat and levels of radioactivity that remain dangerously high for centuries are mainly produced by nuclear weapons and nuclear power, as a result of the splitting of uranium atoms. Whatever their source though, these high-level wastes can remain lethal for tens of thousands of years and are capable of producing cancer in us and birth defects or cancer in our children.

Since Most of Us Aren't Exposed to These Wastes, How Can They Be a Threat to Us?

The government estimates that each of us receives a whole body dose of less than one millirem of radiation a year from all nuclear energy operations, civil and military, including storage of wastes and spent nuclear fuel. This is certainly minor when we consider the average 100 millirems each of us gets from natural sources plus the average 20 millirems we receive annually from x rays, and the four to five millirems we get from both consumer products and weapons fallout. But averages can be very misleading, and in this case the one-millirem figure is particularly deceiving. Averages probably underestimate the threat of radiation from wastes, and mask the fact that some of us get much higher doses than others.

Those of us who live near the mines where uranium fuel for nuclear reactors is excavated, or near radioactive waste dumps, run higher exposure risks, especially in the event of an accident that might leak the wastes into the environment. In addition, we must take into account all the waste that is piling up at reactor sites, many of which are located in heavily populated areas. The reactor-site pools built to hold the waste are rapidly being filled, and the waste will eventually have to be sealed off or transported elsewhere for storage. Either way, an accident, such as one in transport, could expose significant numbers of us — either directly to the radioactive waste, or indirectly to radiation through contamination of the soil, water, and vegetation. And already large quantities of other radioactive waste are being transported on our highways daily.

The magnitude of the waste problem, and the potential harm it presents to us, becomes very clear when we consider the amount of waste generated by just one nuclear reactor, and recognize the fact that every step in the production of nuclear power produces hazardous radioactivity — a fact that the people in numerous western towns are painfully aware of.

Why?

Residents of western towns have been carelessly exposed to uranium tailings, the litter that is left behind when uranium ore is pulverized as the first step in preparing it for use in a nuclear reactor. At one time scientists considered these tailings harmless because of their low level of radioactivity. Later it was learned that in addition to the slightly radioactive dust they give off, the tailings also emit a radioactive gas called radon that can seriously jeopardize a person's health. Although uranium tailings now have to be buried nine feet under ground, for more than twenty years they were simply abandoned in huge piles around uranium mills.

Much of the existing 140 million tons of this radioactive garbage litters the western part of the country, emitting radon gas that is carried by the wind, and contaminating the soil with radioactive substances like radium that are leached out by the rain. Worse yet, as recently as the 1960s the tailings were used in several western states for landfill, and in construction materials for the foundations of homes, schools, and other buildings. In Grand Junction, Colorado, nearly 5000 buildings have been contaminated with the tailings, and apparently many people have begun to pay the price.

In What Way Are These People Suffering?

In 1970, a local pediatrician found an increase in the incidence of cleft palate, cleft lip, and other birth defects in newborn babies in the area. An investigation showed that the children were born to parents who lived in homes built with tailings and that the buildings were giving off high levels of radiation. The University of Colorado Medical Center obtained government funds to study the problem. But a year into the investigation the government cut off the research money for budgetary reasons, concluding that the research wasn't worth pursuing.

At the same time, the Colorado Department of Health calculated that the people who lived in 10 percent of the contaminated homes were being exposed to the radiation equivalent of more than 553 chest x rays a year. And these homeowners aren't the only ones threatened by the tailings.

The Miners, Right?

The miners, as well as people living in the vicinity of the tailing piles, are also threatened. To date it is estimated that more than 230 uranium miners have died as a result of on-the-job radiation exposure. Even if uranium mining were halted immediately, scientists believe an additional 1000 miners could die from accumulated radiation. The greatest risk is to miners who worked before the 1977 ventilation safety standards were enacted to cut their exposure to radon gas. In fact, taking into consideration the latency period for cancer, the Mining Safety Agency claims that thirty to fifty uranium miners who worked prior to 1965 are dying each year from lung cancer. While improved ventilation has reduced the amount of radon in the mines, many critics contend the levels are still too high for the miners' good. What is particularly worrisome is that those who have inhaled the radon gas will retain traces of radioactivity in their bodies for the rest of their lives. This is because radon gas decays to produce a series of radioactive particles that lodge in the lung tissue and emit their own type of nuclear radiations.

What About the People Living Near the Tailings?

According to a Department of Energy report, people who live in close proximity to the piles of tailings double their risk of lung cancer. One huge pile of this radioactive junk is only thirty city blocks from the center of Salt Lake City. In addition, in 1979 the Environmental Protection Agency revealed that mining operations in New Mexico had contaminated the underground water supply of Navajo Indians on a nearby reservation in Crown Point. In all, fifteen water wells have shown radioactivity above the EPA's acceptable safety limit. There have been unconfirmed reports of related birth defects and illness, and the residents have stopped drinking the water, though it is still being used for their livestock. The government is currently conducting studies to determine the health threat posed by the contamination.

Is Anything Being Done About These Radioactive Garbage Heaps?

The government has been after the mill owners to clean up the debris. But in many cases the mills that produced the tailings have long since closed down and many owners cannot be located. Others that have been tracked down are either unwilling or unable to pay the costs. So Congress has passed legislation that will finance most of the cleanup, which is expected to run between $80 to $125 million.

Nonetheless, the mining of uranium continues with about 20,000 tons mined every day and about 37 percent of it coming from Wyoming, New Mexico, Colorado, and Utah. According to the Nuclear Regulatory Commission, the mining and processing of enough uranium to fuel an average modern-day "light water" reactor for one year produces about 4.6 million cubic feet of tailings — enough to fill four White Houses.

On top of all this waste, additional large amounts of low-level waste are produced in the second stage of the nuclear fuel cycle. In this stage several complex processes turn the uranium millings into fuel pellets that produce another 3300 cubic feet of radioactive debris which has to be either buried in trenches or stored in containers underground. This fuel-production waste and the tailings account for about 99 percent of the bulk of the nuclear fuel cycle. Still, it is the one percent of radwaste produced in the generation of nuclear power that by far presents the greatest health threat because of its tremendously higher levels of radioactivity.

Just How Radioactive Is the Spent Fuel?

It is estimated that to dilute the spent uranium fuel from an average modern-day reactor to a safe level one year after it has been removed from the reactor's core it would have to be mixed with all the water in Lake Superior — that's fifteen trillion cubic meters of water. Yet even at that point, as the radioactive nuclides, another name for radioisotopes, continued to decay the waste would still be a serious health hazard for hundreds of thousands of years.

The intense radioactivity of the waste is the result of what happens to the uranium atoms in the reactor core. All modern reactors use roughly 100 tons of uranium as a power source and bombard it with neutrons. Only a relatively small number of these uranium atoms, however, split or fission to produce energy, resulting in radioactive particles that generally remain dangerously radioactive for less than thirty years. A much larger number of uranium atoms, however, do not split. Instead they absorb the neutrons and are transformed into exotic radioactive isotopes or nuclides called actinides. These waste products can remain harmful for hundreds of thousands to millions of years.

That Sounds Like an Exaggeration

Well, it's not. One of these deadly actinides found in nuclear-power radwaste is plutonium-239. It has a half-life of 24,000 years. (A half-life is the time it takes for an element to lose half its radioactivity.) Another waste product, nickel-59 has a half-life of 80,000 years. Technetium-99 has a half-life of 213,000 years, and iodine-129 has a whopping half-life of 15.9 million years. These are just a few of the radioactive isotopes found in high-level radwaste. And what makes them particularly threatening to life is the fact that many of them have a natural affinity for our body's tissues and organs.

This is because numerous radioactive isotopes are chemically similar to natural elements that our bodies assimilate or manufacture. By their very definition, isotopes of any one element are all chemically similar to that element, having the same number of protons in their nucleus and the same number of electrons orbiting around the nucleus. As a result, our bodies can't distinguish between isotopes and their natural cousins. Sometimes this can work to our advantage.

How's That?

If we have a thyroid disorder, for example, a doctor can have us swallow radioactive iodine-131 and using a scanning device he can measure the iodine's radioactive emissions to diagnose the problem. This procedure works because the thyroid absorbs the iodine-131 just as it does natural, nonradioactive iodine.

Unfortunately, the nonmedical radioisotopes in radwastes are much more damaging to our body not only because they are much more potent but often they remain in our body for far longer periods of time. For instance, long-lived radioactive iodine-129, produced by nuclear reprocessing plants and reactors, is deadly. It can accumulate on vegetation where it can easily find its way into cow's milk and then into our body, where it is assimilated the same as natural iodine that exists in food. The result, of course, can be severe damage to our thyroid gland.

Another harmful radioisotope is carbon-14, which emits beta rays. Since carbon is the building block of all life on earth, its radioactive counterpart, carbon-14, can equally damage plants, animals, and humans. Even if a nuclide is not an isotope of an element in our body, the affinity can exist. Plutonium, for instance, has a natural affinity for our bones and also settles in our lungs and lymph nodes, where it deposits its energetic alpha particles. Plutonium-239 is a particularly lethal radwaste. The amount of plutonium contained in the spent fuel from a single reactor after one year of operation would be sufficient to cause fatal lung

cancer in the entire population of the United States if the plutonium were dispersed as fine particles and inhaled.

Even many of the relatively short-lived radioactive nuclides produced in the manufacture of nuclear weapons and the generation
of nuclear power are dangerously compatible with our bodies.

Which of the Short-Lived Radwastes Have an Affinity for the Body?

Tritium, a by-product of nuclear reactors, has a half-life of only
12.3 years. But because it is chemically similar to hydrogen, which
composes all water, it can easily pervade our water supplies and
affect all life forms, producing genetic mutations with its energetic
beta rays.

Strontium-90, which is present in reactor radwaste and in bomb
fallout, has a half-life of roughly thirty years. Strontium is one of
the most dangerous of radwaste materials because it is a chemical
cousin of calcium and has a strong affinity for our bones; it can
easily cause bone cancer and leukemia. There are many more radioactive materials that possess affinities for our body, and the
nuclear fuel cycle and the production of weapons generate vast
amounts of them. When you consider that the average "light-water" reactor has to have between one-third to one-fourth of its
spent fuel rods replaced in its average thirty- to forty-year life span,
you can see the enormity of the problem.

We Stockpile All This Deadly Waste for a Power Source that Is Effective Only for a Few Decades?

That's right. Our current light-water reactors have only a thirty- to
forty-year life expectancy and after that time they themselves become part of the nuclear fuel cycle's waste heap. Most of the parts
of a nuclear power plant like the turbines, bearings, valves,
pumps, and pipes are subject to wear and failure. This is particularly true of the reactor pressure vessel which houses the radioactive reactor core. Due to the continual bombardment of neutrons,
the core eventually becomes brittle and susceptible to failure. As a

result, the plant has to be "decommissioned," and utility executives must decide whether to overhaul the plant or replace it.

So far only a few small reactors have been decommissioned. Many large commercial reactors are scheduled for decommissioning within the next twenty years, and more plants will soon follow. Some may even be forced to shut down prematurely due to economic factors or safety violations or waste storage problems. Whatever the reason, if the life of a nuclear power plant is not to be extended by extensive repairs, or by conversion to another kind of energy facility, the plant — and its radioactive contents — must be disposed of.

This means that from the initial process of uranium mining, through uranium enrichment, plant operations, fuel reprocessing, and eventual dismantlement, the radioactive wastes continue to mount. And what is most shocking is that we are thirty-five years into the nuclear age, and we still lack a permanent storage site for either radwaste from weapons production or the generation of nuclear power.

Why Don't We Have Permanent Storage Sites?

Largely because of shortsightedness on the part of the government, and in particular the Atomic Energy Commission. In the early years of atomic research the AEC, and its successor the Energy Research and Development Administration, failed to incorporate the proper science into their plans for a nuclear future. During the dizzying war years there was not much time to do anything with nuclear wastes from weapons production but store them in steel tanks. When the war was over, the AEC simply continued this practice, and there was little if any pressure on the agency to devise a safer means of storage.

By the mid 1950s, however, the volume of nuclear waste had accumulated to the point where it was necessary to consider other means of disposal. At that time, most of the decisions regarding nuclear technology were in the hands of engineers and physicists who had successfully run the nuclear industry since the war. Many of these people did not regard waste disposal as a significantly new technical challenge, but rather as a routine "garbage disposal" affair, so the problem didn't attract adequate attention and was left unresolved. In fact, the AEC continually ignored warnings and recommendations for improved waste management.

The AEC's complacency is evident in its annual report to Congress in 1959, in which it stated that "waste problems have proved completely manageable . . . there is no reason to believe that the proliferation of wastes will become a limiting factor on future development of atomic energy for peaceful purposes."

Was the AEC Ever Wrong!

It certainly was. The problem of high-level radioactive wastes has become a major barrier to the construction of new reactors and a major rallying point for nuclear energy opponents. In addition, the AEC's lack of foresight has compounded the waste problem, because the early assurances of government and industry officials that everything was under control resulted in relatively little research.

Unfortunately, the Department of Energy fell heir to this short-sightedness, largely ignoring for years the many environmental factors connected with waste disposal. For instance, until recently little attention was given to the problem of the interaction between nuclear wastes and the rock beds in which they are stored, particularly mechanical stress and pressure that develop due to the heat generated by the wastes. As well, the government and the DOE have spent far too little money on the waste disposal issue, especially when you compare the outlay to the money spent for developing the bombs and reactors that generate the waste.

Colorado senator Gary Hart, chairman of the subcommittee on nuclear regulation, has aptly summed up the government's sorry record in declaring that, "If the word *scandal* can be attached to nuclear power, it is that this industry has been permitted to expand for two and a half decades without an acceptable solution for waste disposal." In fact, Hart favors legislation that would ban new plant construction and phase out the operation of existing reactors over a ten-year period thereafter, if the government doesn't formulate a permanent solution by 1985.

Didn't President Carter Propose a Solution for Nuclear Waste?

Yes, early in 1980 he took an important step in that direction, proposing a fifteen-year national effort to develop a safe method for permanently storing the radioactive wastes from nuclear-weapons production and nuclear power plants. The new program calls for four or five years of research and development of various types of geological repository sites, such as salt beds, and improvement of storage techniques, before the government commits itself to a specific approach. Eleven exploratory sites have been chosen, including the government's Nevada nuclear-bomb test site, its Hanford, Washington, dump where military wastes are stored, and several salt domes near the Gulf of Mexico.

The plan calls for selection of a full-scale repository site by about 1985, but whatever type of site is chosen, it won't be ready for the wastes until the mid 1990s. The program is expected to cost at least $740 million in 1980 alone, and it includes funds to clean up nearly fifteen million tons of uranium tailings. Congress will have to approve substantial parts of the program — which amounts to the most expensive national garbage collection and disposal system in the world — including putting up the money for research and development.

Regardless of How Much Is Spent, Can These Deadly Wastes Ever Really Be Safely Isolated?

Recent technological developments suggest that they can, but no one can yet be sure. Over the years, many permanent radwaste solutions have been offered, most notably the rocketing of waste into space, or the burying of it in the Antarctic ice cap.

But these and other options have been abandoned, at least for the time being, for reasons of technology and safety. The idea of allowing the waste's own heat to melt into the Antarctic or Greenland ice caps has been rejected because of the political problems of trying to secure dumping rights, and because an international treaty currently forbids the dumping of nuclear wastes in the Antarctic region. In addition, in order to safely freeze the long-lived wastes, the ice caps would need to remain stable for 100,000 years or more, and no one is willing to bet on that.

Shooting wastes into orbit, or into the sun, or onto other celestial bodies has also been largely ruled out for various reasons, including the fact that an accident during takeoff could contaminate a vast area around the launch site for thousands of years. And even if space disposal were a reality, the sheer volume of waste would prohibit us from rocketing all of it from earth. So scientists have focused their attention on treating the wastes at their sources — for instance, incorporating them into glass or ceramics — and burying them in natural geologic formations such as salt beds, granite, shale, deep-sea sediments, and tufts — volcanic ash solidified by its own heat.

What Are the Advantages of Geological Disposal?

In theory, it appears that geologic media can safely isolate wastes from our environment, or biosphere, for thousands of years or longer. In addition, experts feel that underground burial can be accomplished with straightforward applications of existing technology. For instance, we have the ability to locate sediments, rocks, and minerals within the earth that have remained intact for many millions of years, undisturbed by intruding groundwater, earthquakes, or folds and shifts inside the earth. Furthermore, because of the "Oklo phenomenon," the existence of a "nuclear reactor" in Africa billions of years ago, many scientists have been given reason to believe that radionuclides can be safely placed in a geologic environment for millions of years.

You're Kidding!

No, not too long ago scientists discovered mysterious nuclear-reactor waste products in several places within one of the extensive uranium-ore beds that lie beneath Oklo in the African country of Gabon. These scientists concluded that the wastes were the result of groundwater and chemical conditions several billion years ago that combined to set up a "nuclear reactor" in the ore deposit. The reactor is believed to have sustained a low-power chain reaction for nearly 750,000 years, producing actinide nuclides and radioactive earth. Interestingly, though, the scientists found that both the soil and the wastes remained largely uncontaminated near the reactor area.

Scientists caution, however, that because the exact depth of the natural reactor can't be determined, they don't know for sure that all the radionuclides have stayed in one place. And they are well aware that any contemporary permanent geologic-waste repository would not even closely mimic the chemical conditions in the Oklo ore.

In addition, there are significant obstacles to overcome regarding geological burial, and each of the media being considered has its own peculiar problems. For instance, the presence of brine in salt formations could leach the radwaste, releasing it into the surrounding earth. Also, there are unique engineering problems; and we have no experience with long-term underground storage, and thus no foundation of empirical knowledge upon which to build in order to assure absolute safety.

More important, though, is the fact that geology is a historical rather than a predictive science. So even if an apparently stable geological site is located, there's no guarantee that it will remain stable. For example, suppose a site is chosen that meets some of the major geological criteria for safe storage. That is, that the geological history of the site shows it to be free of groundwater flow, free of earthquake activity, and unlikely to undergo dramatic sea-level changes or glacial erosion for centuries. While such factors would strongly indicate that the site should remain stable long enough for much of the waste to lose its highest radioactivity, unexpected cataclysmic changes cannot be ruled out. It is because of such considerations that a permanent storage site cannot be ready until the mid 1990s.

Can't the Waste Be Reprocessed in the Meantime?

It can, but that would not solve the waste problem. Our government has had the technology to reprocess spent nuclear fuel for over twenty years. For various reasons, however, the process has been used only to recover plutonium from radioactive waste for use in nuclear weapons.

Reprocessing plants can release dangerous radioactive gases into the air. Also, many people don't realize that reprocessing waste for reuse in other reactors increases the total volume of waste one to three times. More important, though, once the plutonium is recovered it is so enriched that a bomb can be produced from a small amount of it. The proliferation of reprocessing technology, therefore, would not only allow many non-nuclear nations to produce nuclear weapons but would also make it possible for terrorists or subversives to fashion homemade bombs, which they could use to kill or maim thousands of people, or even to subject an entire nation to blackmail.

Just such concerns prompted President Carter in April 1977 to defer indefinitely the reprocessing of spent fuel in this country. At the time of his announcement, the president expressed the hope that other nations would follow suit, so that the world could forego the development of an international plutonium economy and the danger that implies. Of course, since many countries are trying to free themselves from dependence on foreign oil, his request was largely ignored.

Where Will All the Waste Be Stored Until a Permanent Solution Is Found?

The waste from nuclear power plants will continue to be stored in vast reactor-site swimming pools, and President Carter recommended that at least one Away-From-Reactor (AFR) government storage facility be established to temporarily hold spent uranium fuel rods from reactors whose cooling pools fill up. Weapons waste currently stored in huge underground steel and concrete tanks, primarily at three government dump sites in Idaho, South Carolina, and Washington, will stay where it is until a permanent disposal method is chosen.

Several thousand canisters of low-level atomic trash from weapons production are lying buried off the Maryland and California coasts, where they most likely will remain. As for other low-level waste, some hospitals and research laboratories have their own storage facilities, though these are usually limited and most don't have storage areas at all; they hire private waste-disposal companies to cart away the garbage or burn it, a practice that is presumed to be safe. However, with radiation standards now in question, many hospitals, universities, and industries are reluctant to burn their radioactive garbage, especially those located in densely populated areas.

While there has been much disagreement concerning effective means for storing radioactive wastes, there is near unanimous consensus that our present storage system is unsafe, especially for high-level wastes, which generate tremendous levels of heat that can easily corrode storage tanks. Liquid high-level wastes present a particularly difficult problem because if corrosion occurs and tanks spring leaks, the radioactive waste can escape into our environment.

Has a Waste Spill Ever Occurred?

Yes, there are many instances. Two of the more prominent ones have occurred in West Valley, New York, and Hanford, Washington. In West Valley, in 1966, Nuclear Fuel Services, a private company, opened a large nuclear fuel reprocessing plant, the first of its kind in the country. After operating for several years, economic factors forced the company to close, but the accumulated waste was never cleared away. Over the years the 600,000 gallons of radioactive liquid and two million cubic feet of radioactive trash have periodically contaminated local water supplies, which in turn have carried radioactivity into Cattaraugus Creek, a tributary of Lake Erie. No one has documented just what effect this contamination has had or will have on the public — or when the waste will be removed. But estimates run as high as $1 billion to decommission the plant and dispose of the wastes, and since the company no longer exists the radwaste problem has been inherited by New York State.

In 1977, the Environmental Protection Agency cited West Valley as a prime example of a commercial burial site that did not live up to its promise: authorization to open the site was based on studies indicating that the radioactive wastes would be perfectly contained for hundreds of years. Instead, the site leaked radiation in just ten years. Even more worrisome, though, are the events that have occurred at the Hanford dump in Washington State.

What's Gone Wrong at the Hanford Dump?

There have been more than twenty reported leaks at the Hanford facility since it was built over thirty years ago, and by 1973 nearly 425,000 gallons of radwaste had seeped into the sandy soil at Hanford. The site, which occupies 570 square miles adjacent to the Columbia River in southeastern Washington, was built during World War II as a giant complex containing a fuel fabrication plant, nuclear reactors, and chemical separation plants for the extraction of plutonium from uranium, the basic ingredient of atomic bombs. Over the years, Hanford's reactors have generated tens of thousands of pounds of weapons grade plutonium and they have produced more than 100 million gallons of highly radioactive waste, which is stored in huge underground tanks.

The tanks were originally composed of a single wall of carbon-steel and concrete. But frequent leaks in the tanks soon forced the Atomic Energy Commission to insist that double-walled containers be used. Currently, the older single-walled tanks contain low-heat-generating and solid waste, while the newer tanks hold all the high-heat-generating waste. The government had initially predicted that, though temporary, the old tanks would provide a safe, effective means of storage for as long as 500 years. Yet by the 1950s, tens of thousands of gallons of radwaste had seeped from the tanks.

Contaminating the Ground?

Yes, and the water as well, with radioactive strontium-90 and cesium-137, which can easily damage our bones and soft tissues. Measurements indicate that the strontium and cesium, plus several longer-lived nuclides, were traveling underground through sediment toward the Columbia River at rates faster than had been expected. Some nuclides, such as tritium, a radioactive form of hydrogen, which is easily incorporated into water and can therefore pervade all life forms, has already been detected along the banks of the river.

All of the leaks have been caused either by corrosion of the carbon-steel tank liners, thermal expansion of the vessels due to overheating, or mechanical defects in the containers. Whatever the cause, by the early 1970s spilled wastes from the Hanford facility totaled nearly a half million gallons.

Not all the radioactive pollution at Hanford has been accidental. For a number of years it was commonplace to discharge plutonium-laden waste into a large ravine. Today, the soil there contains more than 100 kilograms of plutonium, enough to produce nearly a hundred A-bombs, each with a destructive force equal to the bombs dropped on Japan.

Could the Plutonium Explode?

Some scientists feel it might. The Hanford officials had expected the plutonium to spread itself thinly throughout the soil. Instead, it has concentrated itself in the uppermost layers of earth, and there are some fears that a spontaneous chain reaction might occur that could touch off a tremendous nuclear explosion. In 1973 the Atomic Energy Commission warned that the concentration of the plutonium at the bottom of one burial trench might be great enough to cause a low-grade nuclear explosion, and much money was spent to remove the radioactive material.

Many scientists, however, dismiss the possibility of an explosion occurring now, especially a large-scale one. But Soviet émigré scientist Dr. Zhores Medvedev believes that a similar situation involving concentrations of high-level waste in tanks may have been responsible for a widespread nuclear disaster in the southern Ural Mountains in the Soviet Union in the winter of 1957. It is suspected that numerous injuries and deaths occurred from the explosion, which is believed to have contaminated a 360 square mile area forcing the evacuation of residents from nearly thirty villages.

Speculation aside, officials at Hanford no longer discharge the radwaste into the soil, and, in addition, they claim that they have effectively coped with the leaking waste problem, contending there

have been no leaks of radwaste into the soil or water since the mid
1970s.

Are They Telling the Truth?

Apparently not. In 1979 a physicist, Stephen Stalos, who worked
at the Hanford facility for five years, filed a formal complaint with
the Energy Department charging that officials who manage radioactive waste on the site have withheld evidence clearly suggesting
that new leaks had occurred in at least three tanks. According to
Stalos, he attempted to report one of the leaks in June 1977 to
DOE but was told that it was the Energy Department's policy
"that there will be no more leaks," because if the public became
aware of additional leaks such knowledge would adversely affect
the nuclear industry.

The DOE investigated Stalos's later complaint, however, and
has recently concluded that several policies of the Hanford reservation are in need of "wholesale overhaul" and that management's policies have effectively kept publicity about possible tank
leaks to a minimum. The report cited one incident in which a
DOE investigator visited Hanford for two weeks but was never
informed of a pending draft report recommending that six tanks
be added to the list of those leaking radioactive waste. In short,
management has been engaging in an apparent coverup.

Has Anyone Been Harmed by the Leaks or Spills at Hanford or Any Other Dump Site?

So far the spills have caused no obvious harm to animals or residents living nearby. But insufficient time has elapsed to be certain
that the radiation leaked from the wastes will not produce more
cancers and other diseases, birth defects, or illnesses in subsequent
generations. Scientists have calculated, though, what the long-term
effects of similar radwaste spills might be if the radioactive materials eventually work their way into our environment. One such
scenario was reported by the Union of Concerned Scientists in
1980. The report was based on a hypothetical situation where pe-

riodic leaks from tanks resulted in an accumulation of approximately 400,000 gallons of waste, half of which contained strontium-90. The spills were similar in volume to those at Hanford, and calculations were performed to determine what would happen if after a 100-year delay, just 10 percent of the strontium worked its way into the water supply of a city or area of 500,000 people.

What Was the Conclusion?

Following a fifteen-year latency period 1500 to 3000 people in the community would die every year for thirty years from leukemia and bone cancer. Of course this doesn't include the nonfatal diseases that would occur or the genetic mutations or birth defects that might result.

Though such a hypothetical analysis can obviously overestimate or underestimate the damage, it still serves the purpose of illustrating the potential long-term threat to residents who live near significant radwaste spills. Even barring large-scale spills, the health dangers from waste exist. An EPA study claims that during the first 10,000 years of waste storage, when most of the waste is highly radioactive, some people will certainly die from the release of radioactive waste, especially from groundwater leaching. The study didn't define "some," but government and industry acceptance of a small number of deaths is disturbing, and so is the fact that scientists have been consistently underestimating the health dangers posed by radionuclides, both those leaked into the soil as well as those emitted by nuclear power plants into the air.

How Far Off Have Scientists Been?

In the worst case, they have underestimated the dangers of some nuclides by a factor of several thousand, a fact that has only recently come to their attention. At the 1979 meeting of the International Atomic Energy Agency in Vienna, scientists from the University of Heidelberg in West Germany reported that their government and the United States' Nuclear Regulatory Commission have been using incomplete theoretical models to calculate the doses of radiation received by the public from nuclides. This is particularly true for radioactive cobalt. Handbooks for calculating cobalt doses ignore the fact that bacteria in the stomach of livestock and in soil bind cobalt to vitamin B-12 which can then enter the food chain.

German scientists determined that if B-12 binds itself to cobalt in food, 70 percent of the contaminated vitamin would find its way to a person's liver, and remain there with a half-life of 750 days. In simple terms, this means that the radiation dose to the body from cobalt in the environment has been underestimated several thousand times. As a result, scientists at the Vienna meeting emphasized the need to reevaluate radiation-dose models for all nuclides contained in radwastes and emitted by nuclear power plants, and the manner in which nuclides are transferred through the environment and into the food chain.

This new information makes it all the more urgent for our government to devise a permanent means of nuclear waste disposal, and in the meantime to reduce the possibility of future spills.

Is This Being Done?

Yes, among other things, highly radioactive liquid waste must now be evaporated to form solid, salt cakes, which reduce the possibility of seepage. And increasing amounts of lower-level liquid waste are being stored in double-walled containers that allow scientists to detect a leak if it should occur in the inner tank and correct it before effluents escape from the outer tank into the ground. While these are positive steps, permanent storage for high-level waste is years off and even in terms of low-level waste we are running out of places to store the stuff. Basically, many states don't want to supply storage sites, a point that was dramatically underscored in 1970, when three state governors refused to allow major dump sites within their states to accept certain low-level waste.

Why Wouldn't They Accept the Wastes?

Mainly because of sloppy packaging of the waste and fears that inadequate and careless waste-shipping procedures jeopardized the health not only of plant employees, but of the public as well. The dumps involved were at Hanford, Barnwell, South Carolina, and Beatty, Nevada.

Because of its location the Barnwell site has traditionally received most of the East's radwaste shipments. In the spring of 1979, however, South Carolina's governor Richard Riley placed a limit on the amount of wastes that the Barnwell site would accept, refusing such items as radioactive liquids used in laboratory research and contaminated materials discarded after medical diagnostic tests that employ radioisotopes. As a result, haulers were forced to cart such wastes to the Nevada and Washington sites.

In the summer of 1979 the governors of the three states took action to show their dissatisfaction with the Nuclear Regulatory Commission's waste-shipment program. They formed an alliance and petitioned the NRC to take greater responsibility for seeing that wastes were properly packaged and safely shipped. When the NRC did not move quickly, Nevada's governor Robert List temporarily closed the Beatty site. A week later Washington governor Dixy Lee Ray closed the Hanford dump after several particularly dangerous shipments arrived. These shutdowns caused a garbage crisis in the medical community.

Do Hospitals Produce That Much Radioactive Waste?

A large hospital often conducts thousands of radioactive tests and treatments a day, producing enough radwastes to fill 100 thirty-gallon drums every week. So when the dumps closed, many hospitals that depend on the dumps for disposal of their low-level radwastes were faced with huge pile-ups of the hazardous materials. Massachusetts General Hospital in Boston, for one, had to warn physicians not to order any radioactive diagnostic tests that could be delayed. At Columbia University haulers refused to collect radioactive trash, and the laboratories had only two to three weeks of storage space left. Duke University had no more than thirty days worth of storage space remaining, and The Memorial Sloan-Kettering Cancer Center had only a couple of months storage space left. Harvard University claimed it would soon be forced to limit certain kinds of research, and many medical centers curtailed cancer research involving radioisotopes. All of this led to fears and unconfirmed reports that hospitals would be forced to secretly pour certain liquid wastes down sink drains, and that hauling companies would flush "hot" liquids into sewers — dangerous practices that are forbidden by law.

Has the Garbage Crisis Been Resolved?

Only temporarily. In November of 1979, Governor Ray, citing humanitarian concerns, announced the reopening of the Hanford dump. Ray stated that the nation's medical and research centers were engaged in lifesaving work and should not be forced to curtail their efforts. Governor List of Nevada, however, made his summary suspension order of the Nevada site permanent, while Governor Riley announced that South Carolina would continue to refuse shipments normally dumped in other states. The three governors also reached an interstate agreement that requires shippers of wastes to certify that each load meets all federal and state safety requirements, and requests that additional states open dump sites.

But their requests may be in vain. There has been increasing support in Congress for laws providing states with the power to reject waste-disposal sites for both high- and low-level wastes within their own borders. Some states have such high population densities that the presence of dump sites could pose serious health hazards to millions of people. For example, California, the most populous state in the nation, recently halted the construction of the Sun Desert Reactor because an acceptable waste-disposal plan had not been designed.

The increasing confusion and disagreement over whose backyard will house the garbage is compounding the radwaste problem. Already a half dozen federal agencies are vying with growing numbers of state and local governments to decide where the wastes will go. The nuclear-waste plans initiated by President Carter may help alleviate the situation.

How?

Congress is looking into the creation of a state planning council consisting of state governors and legislators and federal bureaucrats who would in turn work with Congress to decide where the nation's radwastes should be stored. In addition, DOE is already moving ahead, expanding its waste-disposal program and increasing funds for studies in various geologic media. In conjunction with the U.S. Geological Survey, DOE is hopeful that its research efforts will resolve the outstanding questions concerning the various geologic disposal sites. While much remains to be done, the government, and particularly DOE, seem to be moving in the right direction at last.

Can DOE Be Trusted? After All, Isn't One of Its Main Objectives the Promotion of Nuclear Power?

That is one of its goals. Though the Nuclear Regulatory Commission ostensibly regulates waste, DOE is largely involved in the formulation of waste-management policies and waste-disposal research, funding more than 60 percent of federal radiation research projects, while at the same time actively promoting nuclear power. The potential for conflict of interest is real, especially since the nuclear industry considers a solution of the radwaste problem essential to the future of nuclear power. One of the major fears is that as a result, DOE will concoct a waste-disposal plan too hastily.

Already the department has been criticized for its rush-to-judgment approach on a waste-disposal pilot project in salt beds near Carlsbad, New Mexico. DOE wanted to develop the site into a permanent one, rather than wait for several alternative sites to be investigated, as President Carter had chosen to do. However, engineers found potentially serious problems with the DOE choice, in particular the existence of pockets of brine, and President Carter canceled the project in lieu of his proposed radwaste policy — although investigation of the site will continue, with the possibility that it might be included in the group of sites from which a final choice will be made.

Recently, the conflict-of-interest fear prompted a special White House task force to recommend that the National Institutes of Health replace the DOE as the head agency for research into the health effects of radiation from all sources, including radwastes.

The task force conceded that much of the research performed under DOE sponsorship has been of high quality and beneficial, but that "the public perceives that a tension exists between DOE's role as the primary sponsor of research into the effects of ionizing radiation and its role as the developer and promoter of nuclear energy." The task force suggested that NIH would be the best candidate to take over radiation–health-related research, since the organization has extensive research facilities, already conducts and supports research into the causes and prevention of disease, and has a staff of hundreds of highly skilled scientists. But while the task force recommendations are sound, some groups maintain that they do not go far enough.

Because They Didn't Recommend that DOE Also Be Stripped of Its Control in the Area of Waste Disposal?

Precisely. Several environmentalists and scientists, including members of the Union of Concerned Scientists (UCS), would like to see a new authority set up, totally independent of DOE, to assume control of radwaste management programs. They feel that because DOE has bungled radwaste policy so badly in the past, and because the department is responsible for the promotion of nuclear energy, it is a poor choice to resolve the radwaste problem. In particular, the UCS has gone on record favoring an independent body run by competent, independent scientists, engineers, and administrators, and an agency head who has no present or past ties to the country's nuclear-energy program. The UCS argues that establishing such an agency would not uncouple the futures of nuclear power and nuclear waste, but rather would eliminate "the political and economic pressures that in themselves pose formidable obstacles to finding and developing a safe, sound solution to the radioactive waste problem."

Wouldn't It Be Politically Difficult to Set Up an Independent Authority to Oversee Disposal Research?

Yes, and costly, too. Yet it could prove to be well worth the effort, and, in the long run, cheaper, considering the inefficiency and mismanagement that have characterized the waste-disposal program in the past. The current odds against establishing such an agency must be considered great, however, not only because of the political and bureaucratic wrangling that characterizes any transfer of authority but also because of the economic conditions in the country. An increasingly cost-conscious and balanced-budget-minded Congress could hardly be expected to fund a new agency. This doesn't mean, though, that the government shouldn't at least investigate the creation of such a body, and push ahead with plans to establish one if studies indicate it would be the most effective way to manage the radwaste problem.

In the meantime, the best we can hope for is that other government agencies, such as NIH, will either assume DOE's functions in the radwaste area or will at least share responsibility for research and development in seeking a radwaste solution. In addition, there is always the hope that if the underpinnings of President Carter's radwaste policy are supported by the Reagan administration, DOE policy will be subjected to much closer scrutiny, ensuring that the best possible means of safe permanent storage will be found.

Radiation from the Sky: Bombs and Fallout

Why Do Bombs Produce Radioactive Fallout?

*B*asically for the same reason that nuclear power plants produce radioactive waste. As the atoms of uranium in a bomb are bombarded by neutrons, many atoms split to produce tremendous amounts of energy and deadly short-lived radioisotopes. Other uranium atoms absorb the neutrons and are transferred into even more deadly long-lived radioisotopes. So when an atomic bomb explodes, it gives off the initial carcinogenic blasts of gamma rays, neutrons, the devastatingly intense heat radiation and shock waves, and all these radioactive wastes as well.

Just seconds after a bomb explodes, these nuclear-fission fragments are thrust into the air along with rock and soil, forming the familiar ominous-looking radioactive mushroom cloud. Many dangerous radioisotopes have extremely short half-lives and thus have little effect on life, other than in the immediate area of detonation. But a substantial number remain hazardous long enough to present a major long-term health hazard.

These particles not only contaminate the immediate vicinity of the blast but can drift on air currents in the upper atmosphere for hundreds to thousands of miles before settling on soil and crops, where they enter our food chain. They can also harm us more directly by entering our mouths, lungs, and open wounds.

Which Fallout Debris Is Most Dangerous?

From a health standpoint, strontium-90, cesium-137, and iodine-131. Strontium because of its affinity for our bones, cesium because it can settle in our soft tissue, and iodine because of its affinity for our thyroid gland. Although plutonium-239 is not created by the fissioning of atomic bomb material, it, too, can be present in fallout, if it is used as a fuel in a bomb — which it often is — and fails to explode. Plutonium, with its half-life of 24,300 years and its chemical similarity to iron, can cause cancer of the liver and bone marrow.

Plutonium is so lethal, in fact, that if one pound of it were to become airborne and if this debris found its way into our lungs, it could kill every man, woman, and child on earth. A rather disturbing thought considering that nearly five tons of plutonium have been released into the atmosphere in the northern hemisphere alone as a result of atmospheric bomb testing by the superpowers between 1953 and 1963. Most of the plutonium has escaped into the upper atmosphere or settled directly on the earth's soil and in its oceans, where it's believed to be responsible for an undetermined number of cancer deaths. Still, of all the various fallout products, it is probably strontium-90 that presents the greatest danger to us because of the ease with which it enters milk products.

Because Strontium Is a Chemical Cousin of Calcium?

Precisely. When a cow eats strontium-laden vegetation, its milk producing organs can't distinguish between the strontium and calcium, so the milk it produces becomes contaminated with radiation. If we drink the milk, the strontium can obviously find its way to our bones along with the calcium. A woman who drinks strontium-contaminated milk or eats strontium-tainted vegetation can contaminate her breast milk, and if she breast-feeds her newborn baby, the radioactive element is passed along to the child. Infants and children are particularly susceptible to radiation damage from strontium, because of the obvious fact that they drink greater quantities of milk than the rest of us, and because their rapid cell growth can propagate any radiation damage more quickly.

Whatever our age, though, strontium-90 produces the same effects. It can concentrate in our skeleton and cause bone tumors and it can settle in our bone marrow and destroy our body's red blood cell producing mechanism causing leukemia. Milk and milk products are basically the only foodstuffs we have to worry about in regard to strontium. Even if we were to eat a contaminated cow's flesh, the threat of radiation damage is negligible since very little strontium accumulates in soft tissue.

As a result of the bomb tests in the 1950s and early 1960s, the average amount of strontium-90 in our bones has increased but there is yet no hard data to prove that it has caused higher incidences of bone cancer or other diseases in the general population.

Do People Who Live in Areas Where the Nuclear Tests Occurred Suffer Radiation Diseases?

That's a different story, and involves not just strontium but all fallout debris. Between 1951 and 1958 — the years in which most atmospheric bomb testing was conducted — residents of Nevada, the primary site of more than eighty U.S. bomb blasts, and adjacent states, were exposed to a high concentration of fallout. It is believed that nearly 4000 infants have died as a result of the fallout, and many of the people who were exposed to the fallout are living examples of its deleterious effects.

One of the largest ongoing fallout studies was prompted by the discovery of an abnormally high rate of leukemia among 3212 military personnel who participated in a 1957 nuclear weapons test known as "Shot Smoky" at Yucca Flats, Nevada. The study was precipitated in 1977 by the case of Paul R. Cooper, an ex-GI who claimed that the acute myelocytic leukemia he had contracted a year earlier was a direct result of radiation he had received in the Smoky test. The U.S. Public Health Service Center for Disease Control in Atlanta took up Cooper's case and launched an investigation of other Smoky participants as well. While researchers concluded that most of the radiation the men received was not from the Smoky blast but from fallout present in the area from previous blasts, they nonetheless found the men to have a leukemia rate double that of the general population. Film badges worn by the men to measure radiation exposure revealed that many of them received 500 millirems — only five times the natural background dose. By present health standards this is regarded as a level of radiation so low that it should not appreciably increase the cancer rate. While it is possible that the film badges did not accurately measure radiation exposure, the fact that so little radiation may have caused the leukemia casts further doubt on just how much radiation, especially from ingested radioisotopes, may be required to induce cancer.

What About Military Personnel Who Took Part in Other Tests?

Currently the National Academy of Sciences is conducting an investigation for the Defense Department into the health of all 250,000 military personnel who participated in the atmospheric atomic bomb tests. The study is expected to take from three to five years to complete, and the agency has set up a toll free telephone number (800-336-3068) to help track down those who were involved in the tests. About 103,000 soldiers are known to have participated at close range — some in trenches only one mile from the center of the blast — and the Defense Department already has compiled preliminary findings on them. The department, though, is keeping the incidences of cancer among these high-risk men a secret — presumably until the study is concluded. But one thing we do know: of twenty-three experimental monkeys that were present at the Smoky blast, half already have contracted some form of cancer.

Are There Other Studies Linking Fallout and Radiation Disease?

There are several. In 1978, Dr. Joseph Lyon, an epidemiologist at the University of Utah Medical Center, headed a study of children living in Utah who were exposed to fallout from the 1950s weapons tests in the Nevada desert. The researchers compared leukemia death rates in parts of Utah that received high levels of fallout with rates in areas that received lower levels. Between 1944 and 1975 the incidence of leukemia did not change significantly in the low-fallout areas. But in the high-fallout regions the leukemia death rate more than doubled between 1959 and 1967. After 1968, it dropped to its preatomic test level. It takes about seven years for radiation-induced leukemia to manifest itself, and, interestingly, the dates of the bomb blasts correspond accordingly to the subsequent increases in leukemia.

Dr. Lyon's research, though, has not gone unchallenged. Critics have attempted to demonstrate that the increased leukemia death rate is actually a statistical fluke, and they stress the fact — an important one — that the quantity of radiation the children were exposed to is not known. Lyon suspects the dose might have been as high as 8000 millirems, compared to the Maximum Permissible Dose of 5000 millirems that radiation workers are allowed to receive in a year.

Perhaps the critics' arguments would carry more weight if it weren't that Lyon's research is substantially supported in part by a 1965 government study showing a high rate of leukemia deaths among people exposed to fallout during the 1950s Nevada weapons tests. There was no way, though, that Lyon could have been familiar with this earlier study when he published his results in the *New England Journal of Medicine,* because the government had kept it a secret.

A Coverup?

Exactly. The study didn't surface until 1979 as a result of the Freedom of Information Act. Although the study, prepared by Edward Weiss, a Public Health Service investigator, did not conclude that there was a direct connection between fallout and leukemia and thyroid disease, it did contain strong evidence suggesting such a link. One portion of the study showed an excessive rate of leukemia deaths between 1950 and 1965 in Utah's Washington and Iron Counties, which are adjacent to Nevada.

During that period, twenty-eight people died of leukemia, compared to nineteen who would have been expected to die according to national averages. Weiss attributed this 47 percent increase to residence in the fallout zone. Another portion of the study, which focused on thyroid disease among children in Nevada, found that seventy children in Washington County had developed thyroid nodules, which can lead to benign or malignant tumors later in life.

After receiving the report, the government debated it for two months. When it finally decided to move, the government issued only a press release on the less controversial, less alarming thyroid study, omitting all references to the leukemia study.

Such Deception Is Shameful!

It is, and it's not an isolated incident. Other recently declassified documents from the Department of Defense and former Atomic Energy Commission show there were coverups from the outset of weapons testing. According to the released material, although government scientists were uncertain of the health effects of fallout, they were convinced that U.S. bomb testing should progress without delay, apparently willing to take the health risk in the interest of military preparedness. Recently disclosed comments by public officials at that time are indicative of the prevailing mood. In 1955, AEC member Willard Libby stated that "People have got to learn to live with the facts of life, and part of the facts of life are fallout." Another AEC member summed up the government's attitude in his statement that "We must let nothing interfere with this series of tests — nothing." Even President Eisenhower had a role in the deception, advising the AEC to keep the public "confused about fission and fusion."

At a Congressional hearing in April 1979 Senator Edward Kennedy concluded that the released notes, memos, and transcripts of the AEC meetings showed "that for over twenty years the Federal government placed the citizens of Utah at risk without their knowledge and without taking proper precautions." The risk was especially grave for the people living in the Mormon towns in the northeast corner of the state.

Why Were These People at Greater Risk than Others?

Because it was the Atomic Energy Commission's policy to test bombs mainly when the wind was blowing away from the big cities in the south — and unfortunately toward the small towns in the northeast. Today, the people in these towns are reporting the highest cancer mortality rates in Nevada.

For instance, the town of St. George, which lies about 145 miles east of Yucca Flats, a major bomb testing area, has won notoriety as "Fallout City." The people of St. George have been exposed to more fallout than was ever measured in any other populated area in peacetime. On May 19, 1953, an accident with the detonation of a bomb dubbed "Dirty Harry" delivered a total dose of 6000 millirems to the residents of St. George in just one day. This is sixty times the natural background exposure the residents of the area accumulate in an entire year. At the time, the AEC's own Maximum Permissible Dose to radiation workers was 3900 millirems accumulated over a thirteen-week period. The seriousness of the St. George fallout is all the more apparent when you consider that today the government accepts a safety standard of only 500 millirems a year for the general public.

St. George also received another mountain of radioactive debris in May of 1953 from the detonation of another bomb code-named "Simon," the nation's first fifty kiloton bomb. Nine hours after the explosion, radiation monitors on the ground reached record levels, and roads leading into the area were closed for hundreds of miles. A total of 250 vehicles that came out of the fallout zone had to be examined, and forty were so "hot" they needed to be totally decontaminated. Throughout this period, the residents of St. George were assured by the government that they were not in any danger. In fact, the AEC sent a public relations team into the area to produce a film of the people of St. George watching the bomb explosions and discussing the fact that the tests did not affect their health or alter their life style.

Time has told a different story, though. Over the last decade residents in St. George and surrounding communities have lost as many as ten family members to cancer.

The fallout from Dirty Harry may provide us with yet another indication of the hazards of fallout. In 1954, about a year after the detonation of the bomb, shooting began in the St. George area on a movie about Genghis Khan called *The Conqueror*. The cast and crew spent thirteen weeks filming the movie in the contaminated zone, and since that time nearly 34 percent of them have developed some form of cancer. Among the dead victims are John Wayne, Susan Hayward, producer-director Dick Powell, and co-star Agnes Moorehead. While there is no hard proof that the tests were responsible, many health experts believe that conclusion is inescapable.

Did Any Fallout from the Bomb Tests Drift Beyond Nevada and the Surrounding States?

It did. Following a nuclear test in 1958 a radioactive cloud drifted southwest and became smog-bound over Los Angeles for four days. In 1953, the radioactivity from a bomb named "Dixie" hung in the air until it reached the northeast, exposing the population of Boston to 1900 millirems of radiation in two days — nineteen times the exposure Bostonians receive from natural sources over a year. The health effects of these exposures have yet to be determined — and probably never will be, since they now are masked by the generally high rate of cancer attributed to stress, diet, smoking, and chemical pollutants.

The possibility, though, that fallout can have damaging effects on people who live far from the initial bomb blast shouldn't be underestimated. In 1962 there was a dramatic increase in birth defects in children born in Alberta, Canada. At first health officials there were at a loss to account for a 78 percent increase over 1959. Many finally came to the conclusion that Soviet bomb tests near the Arctic Circle in 1958 were responsible. The fallout from the bomb blasts is believed to have been deposited on the ground by heavy rainfall in Alberta in 1960 and 1961.

In addition, there has been one claim that drifting fallout has been responsible for a general decline in Scholastic Aptitude Tests scores in this country.

Fallout Can Affect Intelligence?

That's the notion that has been expressed by University of Pittsburgh physicist Dr. Ernest Sternglass. At the 1979 meeting of the American Psychological Association Sternglass reported that recent declines in Scholastic Aptitude Test scores may be the delayed result of atomic bomb testing in Nevada and other states. Sternglass maintains that radioactive fallout in the late 1950s interfered with the development of higher brain functions in children which in turn has contributed, in his words, to a "generation not achieving full intellectual potential." Specifically, Sternglass claims that high levels of iodine-131 in the thyroid gland and of strontium-90 in the pituitary gland have produced a slight mental impairment in children born during the bomb testing years and that this has affected their SAT scores.

Sternglass arrived at this conclusion after studying SAT scores from several test site area states, including California and Utah where test scores fell sharply eighteen years after the weapons tests. In Utah Sternglass found SAT scores plummeted twenty-six points in students born during the atmospheric testing. To support his claim, he says that Utah scores have recently begun to move upward, gaining some ten points in the late 1970s. The jump, he claims, could not be possible if socioeconomic factors alone were responsible for the initial decline. Sternglass also asserts that fallout can be blamed for lower test scores even in states quite a distance from the test area. In New York, for instance, he maintains that low test scores have been caused in part by large amounts of strontium-90 that had been carried east by winds and precipitated by rainfall. To support his thesis, Sternglass points out that Ohio, where SAT scores have remained virtually unchanged, had low rainfall during that same period.

Most educators remain skeptical of Sternglass' statistics and attribute the lower scores to the fact that today more students are taking the test, especially those whose families have not traditionally attended college. A national panel recently arrived at the same conclusion and also blamed increased television viewing, declining motivation, and family troubles for the lower scores. So, in short, there is yet no firm evidence that bomb fallout has adversely affected human intelligence.

Weren't There Any Early Signs that Fallout Was Dangerous?

As early as 1953 there was an indication that above-ground bomb testing was not as safe as the government claimed. In the spring of that year 4200 sheep that had spent the winter forty-eight miles north of the Yucca Flats testing range died. Suspecting that the sheep's deaths were a result of bomb tests, angry ranchers attempted to force the government to compensate them for their loss. But the government insisted that the animals died from natural causes, and a federal judge in effect upheld the government by ruling that the ranchers could not prove a link existed between the deaths and fallout. Now, more than twenty years later, evidence has emerged to support the ranchers' claims.

In 1978, Utah's Governor Scott Matheson, who has had ten members of his own family die of cancer, began scouring the state's archives for clues as to what might be responsible for the high cancer rate in his state as well as the deaths of the sheep. In his search, Matheson uncovered a 1953 U.S. Public Health Service autopsy report on several of the dead sheep that revealed they had a concentration of radioactive iodine-131 in their thyroid glands 1000 times the level permissible for humans. In fact, according to the report the sheep had experienced spontaneous abortions, lambs were stunted at birth, and many animals had lesions consistent with radiation burns found in cattle tested in earlier atomic blasts — facts known to government scientists when they testified at the 1956 trial that the sheep had died of natural causes.

Further support for the ranchers' claims came in the summer of 1979 when Dr. Harold A. Knapp, a former member of the Atomic Energy Commission's Fallout Studies Branch, conducted a private study for the ranchers. Knapp concluded that the sheep did indeed die as a result of fallout from two particular 1953 nuclear tests. The reason for death: the gastrointestinal tracts of the animals were severely irradiated when they ate vegetation contaminated by fission products present in the fallout. The ranchers, armed with the Knapp report, are trying once again in court to fight for reparation from the government.

The Ranchers Seem to Have a Good Case.

They feel they do, and they're not the only ones who are going after the government. Almost thirty years after the first bomb tests, the people of St. George and neighboring communities have filed the single largest lawsuit ever brought to court by U.S. citizens seeking compensation from the federal government. The first phase of the suit, filed in October 1979 in Salt Lake City, Utah, lists 442 plaintiffs, all of them either cancer patients who live, or have lived, downwind of the Yucca Flats proving grounds or who are relatives of cancer victims. Hundreds of other plaintiffs are planning to join the suit, and attorneys acting for the alleged fallout victims are relying on Dr. Lyon's study of leukemia deaths among Utah children to construct their case. In addition, hundreds of other claims have been filed by residents of Utah, Arizona, and Nevada as well as by ex-GIs seeking disability compensation for illnesses attributed to bomb testing. The legal and monetary ramifications could be awesome considering that about 400,000 military personnel and civilians are considered to have been involved one way or another in the atmospheric bomb testing.

These People Can't Really Expect to Prove that Events of Thirty Years Ago Are Responsible for Their Health Problems Today?

Granted, many fallout victims will have problems proving they are just that. Since twenty to thirty years are required for most radiation-induced cancers to develop, the government could rightfully claim that any alleged fallout-induced cancers are the result of any number of environmental carcinogens or other factors. Some simple math regarding the 250,000 military personnel involved in the tests shows how difficult it is to prove the connection between fallout and cancer.

The 250,000 test participants are believed to have been exposed to a minimum of 500 to 1000 millirems of radiation, which statistically can be expected to result in twelve cases of cancer. Now in any average group of 250,000 people in the same age range as the nuclear test participants, 16 percent, or 40,000, can be expected to die of cancer. This means that of the 250,000 men who participated in the bomb tests 40,012 could be expected to develop fatal cancer. So the real question is, How do you isolate the twelve cases caused by bomb testing? This same reasoning of course also applies to the civilians who were exposed.

Additionally, no one is sure just how much radiation each military participant may have received. There are no individual badge readings for thousands of soldiers who participated in the tests because badges often were assigned one to a platoon, under the chancy assumption that all the men would receive the same dose. The Defense Nuclear Agency is attempting the almost impossible task of reconstructing individual doses for military participants, but final results are not expected before 1982; the job is further complicated by the fact that many military personnel records for that period were destroyed by a fire in 1973. In addition, radiation detection techniques at the time of the tests were not as sophisticated as today, and the radiation readings on many film badges and other exposure-measuring devices at the time of the blasts are considered incorrect.

Then the Victims Have a No-Win Cause?

No, because recently there have been encouraging signals from the government. For instance, while only twenty of 400 fallout-related military disability claims had been granted as of August 1979, Congress has ordered the Veterans Administration to reconsider more than 225 others. And, in its search for new evidence on the rejected claims — about half of them from veterans with cancer — the VA plans for the first time to issue guidelines to expedite handling radiation-related claims. Among other issues, under the guidelines the word of a claimant that he was present at a bomb test will be accepted unless military records show otherwise. And when an individual's radiation exposure is not available, it will be presumed that his dose was at the upper limit of the range of exposures according to the records for his unit.

Legislation has also been proposed in Congress to provide blanket compensation for civilian cancer victims in fallout cases. Recently a bill was introduced in the House that would compel the government to accept liability for illnesses resulting from Nevada bomb testing between 1951 and 1958, and in July 1962, when radioactivity was vented from an underground bomb test. The bill covers citizens residing downwind from the test sites in Utah, Nevada, and California. In lieu of medical proof, it assumes that all cancer victims in fallout areas where the cancer rate is above the average for the affected area qualify for compensation. The Senate is also moving to provide some relief for fallout victims. In addition, President Carter appointed a top-level panel to investigate the claims made by Utah, Arizona, and Nevada residents, and he ordered that the long-suppressed Public Health Service studies on leukemia deaths and thyroid nodules among children be reopened and investigated thoroughly.

How Is It that Most People in the Fallout Areas Are Apparently Unharmed, While Others Claim to Have Been Seriously Affected by Fallout?

As mentioned earlier, sensitivity to all kinds of radiation varies considerably from one individual to another — with the possible exception of identical twins. This variability should not be surprising. Just as diet, age, life style, levels of stress, smoking and drinking habits, and histories of disease, as well as inherited immune responses, affect our sensitivity to drugs, bee stings, pollen, viruses, and a host of other ailments, they affect our sensitivity to radiation as well.

And we may in fact react with greater variation to fallout than to other forms of external radiation like x rays and gamma rays. External radiations penetrate our skin and usually produce a more predictable sequence of destructive events in tissues and cells. On the other hand, when we ingest radioactive fallout nuclides like strontium, cesium, iodine, and plutonium, their various migrations throughout our body, accumulations in different organs, and generally different rates of body retention and elimination — which are affected by diet and general health — all contribute to a far more complex picture of radiation damage. Furthermore, simple things like a shaving nick on a man's face or on a woman's leg, or a child's cut finger may provide fallout debris an entry into the body it would otherwise not have had.

As with other types of radiation, sensitivity to fallout differs from one body to another. Our stomach, intestines, bone marrow, spleen, reproductive organs, and tissue of the lymph glands are most easily damaged. So a person who already has a weakness in one of these organs may experience a heightened sensitivity to radiation and more easily develop cancer. Then, too, some of us are fortunate enough to be born with strong immune systems. This enables us not only to better ward off colds, viruses, and diseases, but probably permits our bodies to better repair radiation damage to cells and chromosomes as well. All of these factors — and more — undoubtedly influence how each of us reacts to fallout from atomic bombs, and accounts for why some people have been harmed while others apparently have not.

There is, however, another very important factor in considering the danger of fallout. The amount of harmful fallout debris from a bomb can vary greatly depending on the type of atomic bomb exploded.

Aren't All Atomic Bombs Basically Similar?

While they all give off nuclear radiations, the term "atomic bomb" actually applies to two very different weapons — the "fission bomb," which is better known as the A-bomb, and the "fusion bomb," more familiarly referred to as the hydrogen, H-bomb, or thermonuclear bomb because of the tremendous amount of heat required to explode it. In terms of fallout debris, the A-bomb is considered "dirty" while the H-bomb is comparatively "clean." That's because the splitting or fissioning of atoms in an A-bomb produces more radioactive by-products than the joining or fusing of atoms in an H-bomb. It's quite simple to understand why.

Atoms are not easily split. Larger atoms, though, can be divided more simply than smaller atoms and that's why the two largest atoms, uranium and plutonium, are used as the primary ingredients of fission bombs. When these large atoms shatter, causing the bomb to explode, they produce nuclear fission fragments, like strontium and cesium, that compose the fallout.

As you might have guessed, if large atoms are the easiest to split, the small atoms are the easiest to fuse together, which explains why "H" or fusion bombs are made of the lightest element, hydrogen. During the fusion process there is no shattering of atoms, so when a hydrogen bomb explodes, theoretically it should produce no fallout. In reality, this is not the case, since fissionable uranium or plutonium is used to detonate a hydrogen bomb. Hydrogen bombs, therefore, are clean only in comparison to fission bombs. All the 235 bombs tested above ground during the Cold War years produced fallout. Though a fusion bomb is cleaner than a fission bomb, it's just as lethal. In fact, an H-bomb can even be deadlier, in terms of the radiation it emits.

Why Is That?

In principle, there is no limit to the amount of hydrogen that can be used to fashion an H-bomb, while there is a limit to the amount of uranium or plutonium that can safely be used to make a fission bomb. The reason for the difference is fairly simple. The fusion process is set in motion by heating hydrogen atoms to millions of degrees. Any amount of hydrogen fuel can be included without danger of a premature explosion, since an explosion cannot occur until the heat is applied by some appropriate means.

This is not the case with a fission bomb. If you concentrate about ten pounds of uranium-235, for instance, into a mass about the size of a grapefruit, the density of the mass will stimulate the uranium atoms to split on their own, causing them, in scientific terms, to "go supercritical." As the atoms split, they release neutrons which in turn split other atoms. In short, this releasing-splitting process sets up a chain reaction that causes the fissionable uranium to explode. Once this supercritical state is achieved, nothing can prevent the bomb from exploding. For this reason, when making an atomic bomb two or more *sub*critical masses of uranium are used and slammed together at the time of detonation to form a *super*critical mass.

How Can Specks of Matter as Infinitesimally Small as Atoms Yield So Much Power?

It all has to do with the interplay of matter and energy. An atom's nucleus is held together by a tremendous force called *binding energy,* and the nucleus must be bombarded with neutrons before it will divide. When the atom splits, it breaks into two fragments of almost equal mass. The total mass of these two new atoms, however, weighs less than the mass of the original atom, because matter is lost in the fission process. According to the laws of nature, the lost matter must be converted into energy. Only an infinitesimally small amount of lost matter (m) is required to produce incredible amounts of energy (E), as Einstein predicted in his famous equation $E = mc^2$.

The same laws of nature apply in the fusion process. When two atoms of hydrogen are combined to fashion a single new atom of helium, the new helium atom weighs slightly less than the total mass of the individual hydrogen atoms. Here, too, the lost matter is converted into a vast amount of energy.

What About the Neutron Bomb? Does It Produce Much Fallout?

Surprisingly, very little radioactive debris. The "neutron bomb" is a variation of the hydrogen bomb, and is actually no bomb at all but rather a miniature thermonuclear warhead for tactical missiles and artillery shells, which kills with an instantaneous burst of penetrating neutrons rather than with the searing blast of an ordinary atomic weapon.

The concept of the neutron bomb was developed by California physicist Dr. Samuel Cohen in 1958. Cohen realized that since the fusion process produces mostly streams of neutrons and relatively little radioactive fallout, it would be possible to construct a remarkably "clean" bomb simply by detonating a small fusion, or hydrogen, bomb that is triggered by a very tiny — and hence not very dirty — fission, or atomic, bomb.

When the bomb explodes, its hydrogen core undergoes fusion and releases billions of neutrons. These are ten times as deadly as x rays. Detonated at an altitude of 3000 feet, little if any of the blast would reach the ground. But neutrons would flash through enemy tanks, barracks, and buildings, inflicting severe radiation damage on people. Depending on their distance from ground zero — and thus the dose received — these victims would experience vomiting, hemorrhaging, loss of control of bodily functions, and death. People out-of-doors within a 1000-foot radius of the blast would be disabled in minutes and die within two days; at 4000 feet, people would be disabled in hours and die in six days. The longest a victim might survive would be several agonizing weeks. Since neutrons travel a limited distance, people more than a few thousand feet from the blast could safely shield themselves in basements or bunkers.

If neutron bombs were strategically exploded over a given area, homes, buildings, and agricultural crops could remain undamaged, and only people would be destroyed. And since there is virtually no fallout — and only minimal radioactivity induced in irradiated materials that absorb neutrons — the bombed region would be reinhabitable in a matter of weeks or months.

Then Why Has There Been So Much Opposition to the Development and Deployment of Neutron Bombs?

There are two major reasons. Opponents of the neutron bomb feel that because it would inflict limited damage, the use of nuclear weapons would become thinkable, and this could possibly lead to a nuclear holocaust. Critics also contend that the bomb is immoral, since it destroys only people and not buildings. This is a point the Soviets capitalized on when they drummed up their propaganda campaign to discourage the United States from developing the weapon. Their message charged that the weapon was typically a capitalist bomb because it spared property at the expense of people.

On the other hand, proponents of the neutron bomb claim that it is precisely the kind of weapon that would deter a Soviet attack on Western Europe. And if an attack should occur, the neutron bomb would offer the best and safest way to repel the Soviets. Military strategists have created several scenarios in which neutron bombs are the most potent weapons. For instance, if the Warsaw Pact forces attacked West Germany, the country could be overrun in a few days unless conventional nuclear weapons were used. Yet to do so, according to one NATO estimate, even a moderate barrage would kill at least five million West Germans. In short, to save West Germany the Allies might have to destroy it. Neutron bombs, on the other hand, would allow selective strikes at Soviet forces, sparing innocent civilians from the destructive force of conventional bombs, deadly radiation, and hazardous levels of long-lived fallout.

In 1979, President Carter decided to postpone development of the neutron bomb, but to modernize existing weapons so they could accommodate neutron warheads if these were constructed in the future. Recent Soviet behavior, particularly the invasion of Afghanistan, has spurred many military officials to push ahead with development of neutron-weapon technology.

Getting Back to Fallout, Is Any from the Bomb Tests Still in the Atmosphere?

Most of the radioactive particles from all atmospheric tests have settled to earth. This is particularly true for the heaviest nuclide plutonium, according to a team of scientists who have examined plutonium levels in Antarctic ice. Since snowflakes trap all sorts of pollutants, and since Antarctic snow does not melt, the snow is believed to provide a frozen record of what has been washed out of the air.

In January 1977, scientists sought to examine this record by digging a five-meter-deep hole in the Antarctic ice, sampling ice at different depths, then melting and analyzing it for evidence of plutonium. They found significant amounts of plutonium at levels they believed corresponded to snow that fell during 1955. They also discovered that the concentrations continued to increase during the bomb testing for a number of years thereafter, but began to diminish when most of the tests were halted. The scientists concluded that in 1979 almost all the plutonium fallout from bomb testing had been removed from the atmosphere. And because of the 1963 Nuclear Test Ban Treaty, under which most of the major powers, including the U.S. and U.S.S.R., agreed to halt atmospheric testing, the atmosphere should stay relatively free of fallout, save for the infrequent testing of nuclear weapons by countries like France and China, who have yet to sign the treaty. The treaty also bans the signatories from weapons testing in space and underwater. Testing at 1000 feet underground is permitted, however, since the fallout debris is confined to a relatively small area.

Fallout May No Longer Present an Immediate Threat, but What About Descendants of Fallout Victims?

Dr. Linus Pauling, who won a Nobel Peace Prize for his fight to halt atmospheric testing, contended that the fallout from a single year of bomb tests would "ultimately be responsible for the birth of 230,000 seriously defective children and also for 420,000 embryonic and neonatal deaths." Pauling made that dire prediction in 1958. While there is evidence that several thousand children born *during* the decade of testing have experienced higher rates of leukemia and death, there is no hard proof that the children of fallout victims, or even their grandchildren, are suffering radiation diseases. Yet it would be premature to assume that there has been no inherited fallout damage since such effects are very difficult to estimate in a population.

Because of the Latency Period?

Yes, as well as for other complicating factors. For radiation to cause genetically inherited damage it must reach the gonads — the ovaries in women, the testes in men. However, not all bomb radiations affect the sex organs. Since alpha particles — and to a lesser extent beta particles — are capable of penetrating only a few layers of our skin, there is little likelihood that this kind of fallout radiation would influence genetic mutations. Even if we were to ingest most of the fallout elements that emit alpha and beta particles, there would be little threat of sex-cell damage, since most of these elements are not biologically compatible with tissues near the gonads. For example, beta-emitting strontium-90 accumulates largely in the bones.

Cesium-137, which emits beta particles and the far more penetrating gamma rays, is deposited fairly uniformly throughout the body tissue. Its danger to the gonads is somewhat lessened by the fact that it is eliminated from the body within a few months. This leaves plutonium-239 and mainly gamma rays and neutrons as the primary fallout radiations that may cause genetic defects that can be passed on to future generations.

There are many gamma-ray-emitting fallout isotopes that concentrate to varying degrees in or near the gonads: iodine-131, cobalt-60, krypton-85, iron-59, zinc-65, to mention a few. Plutonium also concentrates near the gonads. Nevertheless, it is extremely difficult to estimate the collective threat posed by so many isotopes, whose radiation strengths differ greatly. So in the matter of determining what the inheritable consequences of fallout might be, we're forced to look, as are the geneticists, to the survivors of the atomic blasts in Japan.

Do the Survivors of Hiroshima and Nagasaki Provide Us with Any Answers?

The results of recent studies of the children of the Japanese A-bomb survivors are somewhat surprising, and scientists are still uncertain about interpreting the evidence. One thing is certain: the long-held fear that the aftereffects of bomb radiation might produce a race of "monsters" is being proved groundless. The subtler effects of radiation, however, are harder to detect.

The atomic bombs dropped on Hiroshima and Nagasaki in August 1945 killed about 214,000 people instantly, while another 111,000 are believed to have died later of injuries, radiation sickness, and disease. In the wake of the explosions there were predictions that future generations of Japanese would suffer hereditary damage, namely, malformations and cancer from the initial bomb radiations and the subsequent fallout, and several studies were undertaken to detect such effects.

The studies fall into three main categories: *In utero,* involving fetuses in the womb at the time of the blasts; *postbomb,* involving malformations among children conceived after the bombings; and more recently, surveys of children born to A-bomb survivors since 1946. The *in utero* group, numbering fewer than 2000 fetuses because of low conception rates in wartime Japan and the high number of direct bomb casualties, show clear evidence of radiation damage. Children whose mothers received large doses of either neutrons or gamma radiation were shorter, weighed less, had abnormally smaller chests, and suffered brain damage and retardation. Mortality rates, especially during infancy, were higher and increased with the radiation dose received by the mothers.

The postbomb conception study was extremely detailed. Between late 1947 and 1953, researchers attempted to examine the termination of every pregnancy by birth, stillbirth, or abortion in the two cities. About 75,000 newborn children were examined in parents' homes by traveling teams of doctors and nurses. Of these, 20,000 were given second examinations when they were eight to ten months old. Today, these children of parents exposed to atomic bomb radiation suffer no greater incidence of disease than other Japanese young adults.

In the third study, begun in 1958, 53,000 children born between 1946 and 1958 were divided into three groups: those with one or both parents within 1.2 miles of ground-zero — the explosion center — and thus exposed to heavy radiation; those with parents beyond 1.5 miles from ground-zero; and children whose parents were not in Hiroshima or Nagasaki during the bombing. In a third analysis of these three groups, completed in 1975, scientists found that death and cancer rates among survivors' offspring were almost exactly the same as among children of unexposed parents.

Then Are the Fears About Hereditary Deformities from Nuclear War Unfounded?

Not necessarily, since genetic mutations may skip one or more generations. If, for instance, a damaged gene is "dominant" — such as the gene which produces dwarfism — it will appear in the next generation even if it is contributed by only one parent. But if an altered gene is "recessive" — such as the genes for mental retardation and a certain kind of diabetes — its effect will appear only when such defective genes are contributed by both parents. It so happens that many radiation-induced defects, such as impaired mental functions, physical deformities, and reduced resistance to disease, are attributed to recessive genes and may require two, three, or more generations to appear — if at all. For this reason, scientists are working on longer-range genetic studies of both the Japanese A-bomb survivors and descendants of victims of fallout from U.S. nuclear bomb tests. So the threat of a substantial number of inherited genetic mutations is a real possibility from either nuclear bomb testing or nuclear war.

V   Nuclear Power

Should It Have a Place
in Our Future?

*Doesn't the Primary Concern over Nuclear Power Plants Focus
on a Reactor Accident Rather Than the Pile-Up of Deadly
Waste?*

The possibility of a reactor accident is definitely the major concern to most people — especially in the wake of Three Mile Island. A popular misconception and fear, though, is that the worst kind of nuclear accident would result from a power plant explosion. But a reactor is not a bomb. The "light water" reactors which currently comprise the bulk of the nearly seventy operating reactors in this country use uranium fuel that has a readily fissionable uranium content of only 3 percent, and this uranium, no matter how large the amount, cannot produce a bomblike explosion.

Reactors can, however, meltdown and produce what is commonly referred to as a "worst case" nuclear accident. There are many scenarios for such an event, but the one most commonly referred to is the loss-of-coolant accident, or LOCA, which is what occurred both at Three Mile Island and in the popular movie *The China Syndrome.*

What Happens in a Loss-of-Coolant Accident?

To answer that, it is crucial to understand the dual role that a coolant, such as water, plays in a reactor. The reactor core in commercial light-water reactors is surrounded by water. When fission occurs, the heat produced by the splitting atoms turns a portion of the water to steam, which in turn heats other water to steam. The steam drives the turbines that generate electricity.

The water also serves as the reactor coolant. The splitting of uranium atoms produces temperatures in excess of 4000 degrees Fahrenheit at the reactor core. If for any reason the coolant water were lost, or kept from absorbing this excess heat, the temperature of the accumulated radioactive material would continue to rise, overheating the core and causing a rapid "meltdown" of the fuel. This radioactive mass of molten material could possibly rupture the reactor and melt its way deep into the earth, hence the name *China Syndrome;* in reality, however, at its deepest the material would join with the molten lava which lies at the earth's core.

The molten mass, though, is not what's to be feared. Rather it is the radioactive steam cloud that would be produced that could blow the roof off the reactor containment building. Hundreds of thousands of people could be killed by the radioactive cloud, and there would likely be billions of dollars worth of property damage.

What's the Chance of a Meltdown?

Nuclear energy proponents argue that one is practically impossible. They point to twenty years of reactor operation without a meltdown as proof, and claim that in the worst case a meltdown would be prevented by the ECCS, equipment emergency core cooling system, which would automatically flood the core area and lower the temperature. Proponents also claim that even if there were a meltdown, the release of radioactivity would be retarded by the reactor vessel's thick walls. And they say if the reactor were to melt through the vessel, the radioactive material would still be inside the containment building, which is equipped with many devices to isolate the radioactive elements and prevent them from escaping into the environment.

What's more, nuclear power advocates claim that only if very high pressure were to build up inside the reactor building could large amounts of radioactivity be released, and the probability of that happening they argue is extremely small — even if a meltdown should occur. To support their low-risk claim, proponents have traditionally cited risk-analysis studies, the most famous being the Rasmussen report.

What's the Rasmussen Report?

It is the result of a reactor-safety study commissioned by the Atomic Energy Commission to estimate the probability of various types of reactor accidents. The report, published in 1975, was named after the study-group chairman Norman C. Rasmussen of the Massachusetts Institute of Technology.

In drafting their report the commission members used mathematical methods that have been applied for several years in Britain to predict the probability of industrial accidents. They arrived at the conclusion that the probability of a serious reactor accident, such as a meltdown, involving 1000 or more fatalities, is one in 1,000,000 years, given the existence of 100 reactors. The report estimates that such an accident could result in 300 deaths from radiation sickness, and another 5000 deaths and 3000 genetic casualties over a thirty-year period.

These statistics, however, are based on the assumption that in the event of an accident 50 percent of the population living within twenty miles of the reactor could be evacuated successfully within two hours, and that 90 percent could be evacuated within eight hours. If evacuation efforts weren't successful, the study concludes that deaths and injuries might triple.

While such numbers are alarming, according to the Rasmussen report as well as numerous other risk studies, the probability of a catastrophe is practically nil. In short, you have a greater chance of being killed by lightning than by a nuclear accident.

Are Such Low-Risk Estimates Valid?

No, they are not, and since the publication of the Rasmussen report there is a growing body of opinion that disputes the commission's findings. For one thing, in the spring of 1979 a panel of radiation scientists from the National Academy of Sciences issued a paper entitled "Risks Associated with Nuclear Power," in which they concluded that the so-called risk experts do not know enough about the chances for a nuclear accident to speak with confidence about the overall hazards of nuclear power. Even barring a major accident, the report estimated — based on the current projection levels of nuclear power — that 2000 Americans will die from cancer as a result of nuclear power generation between 1975 and the year 2000.

Also in 1979, the Nuclear Regulatory Commission, which oversees the licensing and operation of the nation's nuclear power plants and is certainly no foe of nuclear power, announced that it could not support major conclusions reached by the Rasmussen report, in effect, rendering the report useless as a weapon of nuclear power proponents.

But what really dealt the Rasmussen report its biggest blow were the events of March 28, 1979, at Three Mile Island (TMI), Pennsylvania, the site of the nation's worst nuclear accident to date. TMI proved that we cannot rely on computer predictions alone to determine reactor safety. For example, according to computer calculations, in the worst possible reactor accident no more than 1 percent of the metal in the reactor core would react chemically with water to produce hydrogen. In fact at TMI — which is not considered a major nuclear accident — between 25 to 30 percent of the metal reacted chemically to produce hydrogen, and it's this build-up which could have led to a powerful explosion inside the containment building. TMI also demonstrated the unreliability of risk studies in another way: though the Rasmussen report assumed that human failure rates are higher in a nuclear plant than technical or mechanical failures, TMI showed just how unpredictable and important the human factor is.

What Is the Probability of Human Error?

It's very hard to calculate, but a vital consideration. An accident may be estimated to have an extremely low probability, but people can intervene in such a way as to make the accident inevitable.

One of the most significant findings of the Kemeny Commission, the panel ordered by President Carter to investigate the accident at TMI, was that though mechanical failures were involved — namely, a malfunctioning valve which spilled water out of the reactor's coolant system for two hours — the accident was basically caused by a failure of people. An NRC report on the accident drew the same conclusion.

For instance, at the peak of the crisis, control-room operators mistakenly took the cooling system out of action by cutting back the flow of water. As a result the reactor core became uncovered for an extended period, causing dangerous fuel melting. The Kemeny group reported that if the throttle to cut the water had not been erroneously pulled, core damage would have been prevented.

In addition, on the first day of the crisis a small explosion of hydrogen gas was not reported to NRC investigators because controllers at the plant had apparently rolled up a crucial recorder sheet inadvertently, concealing important data. And according to the NRC, a key element in the cooling system — three auxiliary feedwater pumps — had been taken out of commission before the accident in violation of federal requirements. Had they been left in operation, NRC officials speculate there would have been a completely different, much less serious, outcome.

These are only a few of the numerous human errors that combined with mechanical failures to almost spell disaster at TMI. And it wasn't the first time that such errors almost touched off a major nuclear accident.

Where Else Did It Happen?

At the Tennessee Valley Authority's Browns Ferry Reactor in Alabama. In March 1975, two technicians there used a lighted candle instead of a smoke bomb to search for air leaks. Flames from the candle set fire to insulation cables that were not sufficiently fire-retardant, and the fire spread quickly through the wiring, knocking out the reactor controls. As pressure in the reactor mounted, the cooling system failed, and the water level dropped dangerously low. It took more than ten hours to repair valves and connect new cables to a water pump to prevent a meltdown. The radiation, however, was contained, and today such cables must have more fire-retardent insulation.

A nuclear accident at the government's Idaho Falls test site in 1961 has also been linked to human behavior. In January of that year, an experimental reactor went out of control killing three technicians — one of whom became impaled by a reactor rod. The victims' bodies were so badly contaminated that they had to be buried along with nuclear waste. Initially, the AEC asserted that one worker might have accidentally pulled the rod out too far. But recent evidence has surfaced alleging that one of the victims may have deliberately sabotaged the plant because of a love affair involving his wife and another technician.

These are just two of the reactor accidents attributed to human error; there are numerous others that have resulted from technical problems and malfunctioning equipment. Looking at the record it's no surprise that an accident like Three Mile Island occurred. In fact, according to the Kemeny commissioners it was bound to happen sooner or later.

An Accident like Three Mile Island Was Inevitable?

Exactly. After six months of investigating all the factors surrounding and contributing to the accident at Three Mile Island, the twelve-member commission uncovered hundreds of errors, deficiencies, and problems relating specifically to the accident and nuclear power plant operations in general. They concluded that because of deficiencies in design of nuclear power equipment, inadequate training of operators, confusing procedures, and a complacency that pervaded the industry an accident like Three Mile Island was eventually inevitable.

The Kemeny group laid much of the blame for the sorry state of affairs on the NRC and charged that the agency was "unable to provide an acceptable level of safety in nuclear power." The members maintained that the NRC was a headless organization that lacked a sense of direction and was incapable of administering safety procedures on a day-to-day basis. The commissioners were no less critical of the thinking of NRC officials. In their hearings, the committee members heard repeatedly about the flawed "mindset" of NRC and nuclear industry officials and concluded that such charges were indeed accurate.

How Is the Thinking of NRC Officials Flawed?

According to the Kemeny group, the NRC and the portion of the industry that it examined seem "to be hypnotized by their equipment." The commissioners warned that even though much of the equipment was good, this is a dangerous attitude that must be changed to admit that "nuclear power by its very nature is potentially dangerous, and . . . one must continually question whether the safeguards already in place are sufficient to prevent major accidents."

The commission also found that the NRC and the industry had a tendency to deal with so-called large-break accidents, "the very scary" ones, rather than small-scale ones. In a large-break accident operators are not very much involved since sophisticated safety equipment reacts automatically within seconds. On the other hand, in a small accident like the one at TMI the operators must cope almost singlehandedly. Because of the emphasis on large-break accidents, the commission discovered that the TMI operators' training and depth of understanding "left them unprepared to deal with something as confusing as the circumstances in which they found themselves."

The Kemeny inquiry also found the NRC's interests to be tied too closely to those of the nuclear industry. When the NRC was separated from the old Atomic Energy Commission, the purpose of the split was to separate the regulators from those who were promoting the peaceful uses of nuclear energy. But the committee members claim they saw evidence that the NRC is still influenced by some of the old promotional philosophy, and that the agency sometimes erred on the side of "industry convenience rather than carrying out its primary mission of assuring safety."

What Did the Commission Propose to Correct the Situation?

To abolish the NRC and replace it with an independent executive branch agency ruled by a single administrator instead of the five-member NRC type panel, and to establish a presidential oversight committee to monitor the performance of both the new federal agency and the private companies that design, construct, and operate nuclear reactors. These, however, were only two of several major recommendations made by the commission to improve reactor safety. Others include a call for the following:

• Periodic renewal of operating licenses to be conducted once every four to five years and subject to public hearings.

• The siting of all future nuclear power plants away from large population centers where many are now located. To use New York as an example, midtown Manhattan lies only thirty miles from the Indian Point nuclear power reactor complex. Nearly twenty million people live within a fifty-mile radius of the plant's two operating reactors, which have had their share of safety violations.

• Federal approval of state and local emergency plans in the event of an accident before any utility is granted an operating license. Most states with nuclear facilities have no such federally approved plan. Worse yet, some communities have no plan at all; for instance, there is no contingency plan to evacuate New Yorkers in the event of a TMI-type situation or worse.

• Numerous changes in the recruiting and training of power plant operating personnel and in the instruments available to help operators understand the condition of reactors and to help them measure accidental releases of radiation. This includes revamping the warning display panels in control rooms. At the outset of the TMI accident over 100 alarms began ringing simultaneously, greatly adding to the control-room confusion.

With All the Problems They Found, Why Didn't the Commission Members Recommend a Moratorium on the Construction of New Power Plants — at Least Temporarily?

The commission came close to doing just that — a moratorium until President Carter and Congress had the opportunity to review their recommendations.

Six of the commission members favored such a proposal, and of these John Kemeny, Dartmouth College president, who headed the investigation, and three others, wanted an even stronger proposal. But the moratorium motion did not pass because of a procedural rule requiring support of seven members — a majority — for adoption of a formal recommendation. Instead, the commission members steered a middle course, claiming that its findings alone did not require "that nuclear power is inherently too dangerous to permit it to continue and expand as a form of power generation." However, they cautioned "neither do they [the findings] suggest that the nation should move forward aggressively to develop additional nuclear power." If we "want to confront the risks that are inherently associated with nuclear power," they added, "fundamental changes are necessary if those risks are to be kept within tolerable limits."

Did the Commission Determine How Close Three Mile Island Came to a Meltdown?

The commission didn't estimate how close TMI's reactor came to a meltdown. Though, by its calculations, the panel concluded that even if a meltdown had occurred it was highly probable that the containment building, in combination with the hard rock on which it is built, would have prevented large amounts of radioactivity from escaping. The members warned, however, that such results "derived from very careful calculations which hold only in so far as our assumptions are valid, and we cannot be absolutely certain of these results."

A report, though, sponsored by the NRC and made public six months after the Kemeny report of November 1979 concluded that the reactor in unit two came closer to melting down than anyone realized — even officials of Metropolitan Edison, the utility that owns the two TMI reactors. According to the report, prepared by an investigative team headed by Washington lawyer Mitchell Rogovin, if a shift foreman had not spotted the stuck-open valve that was leaking reactor coolant and had not blocked it off, "within thirty to sixty minutes a substantial amount of reactor fuel would have begun to melt down — requiring at least the precautionary evacuation of thousands of people living near the plant, and potentially serious public health and safety consequences for the immediate area."

In fact, the government considered the possibility of a meltdown real enough to rush supplies of an iodine-blocking medicine to Pennsylvania.

You Mean There's an Antidote for Radiation?

There is at least one antidote — potassium iodide — for radioactive iodine-131, one of the most hazardous radioactive isotopes that would be released in a major nuclear accident.

Iodine-131 can easily find its way to the thyroid gland and cause cancer. Potassium iodide prevents this from happening by saturating the thyroid, preventing the absorption of the radioactive isotope.

Even before TMI, the FDA, at the urging of several concerned scientists, began stockpiling large amounts of potassium iodide to protect hundreds of thousands of people if necessary. One of the scientists who pressed the FDA to take this action, Princeton physicist Frank von Hippel, calculates that people living as far as 200 miles downwind of a major iodine leak would need protection. Von Hippel, along with other scientists, has advocated a virtual nationwide distribution of potassium iodide.

Some states aren't waiting for the federal government to take action. They are developing their own plan to store potassium iodide pellets for emergency use. Of course, potassium iodide is only effective against radioactive forms of iodine and must be taken before irradiation occurs. For people who would become irradiated without such protection, or from other radioisotopes, little can be done, although doctors hope that bone marrow transplants might eventually prove a cure.

Is There Evidence that Bone-Marrow Transplants Are Effective Against Radiation Illness?

Researchers have found that mice injected with bone marrow following irradiation have survived what would otherwise be lethal doses of x rays. The bone-marrow cells work by migrating to the bones of the irradiated animals where they multiply, compensating for cells killed by radiation. Bone marrow is essential to life because it produces the red blood cells that carry oxygen to tissues throughout our body. Interestingly, in numerous cases bone-marrow transplants have been successfully performed between animals of different species. This is possible because irradiated animals, including humans, lose much of their ability to reject foreign tissue.

Transplants of bone marrow have already proved to be of some value in combating the effects of large doses of radiation in humans. In 1958 a nuclear reactor accident in Yugoslavia exposed five people to near lethal doses of neutrons and gamma rays. They were immediately flown to Paris, where four victims received human bone-marrow transfusions and benefited considerably from the treatment.

In time scientists are hoping to culture bone marrow in laboratories and transfuse it into victims suffering from radiation sickness. Scientists are also working toward transplanting vital organs irreversibly damaged by high radiation doses. Of course that breakthrough is many years in the future, and, even if such treatment proves effective, for obvious reasons it would not be practical on a wide scale.

In Terms of Radiation Exposure, Is There Any Indication that the People Around Three Mile Island Have Been Harmed?

In the wake of the accident many people in the Harrisburg area complained of rashes, gastrointestinal illnesses, headaches, and a peculiar metallic taste in their mouths. Federal and state officials have largely attributed these symptoms to the flu or psychological factors, namely the fear of living in the shadow of the nuclear plant. However, because radiation was emitted into the environment, the government, through HEW (now the Department of Health and Human Services), launched a long-range study of all families, encompassing some 50,000 people possibly affected by the accident. While retrospective studies have been made of Japanese A-bomb survivors, this study is the first ever undertaken to examine the effects of low-level radiation from essentially the time of exposure. Because radiation-induced illnesses, like most environmentally caused diseases, require years to develop, epidemiologists say that they won't be able to give the people of central Pennsylvania a definite score card for many years to come — if ever.

Were the Workers at Three Mile Island Harmed?

So far no serious ill effects have been reported, but the government study is also exploring the problems that any radiation exposure may have caused the workers. Nuclear power plant workers, though, are allowed an average of 5000 millirems of radiation a year compared with 500 millirems for the rest of us. On the average they receive an annual dose of 760 millirems a year, much less than uranium miners who provide fuel for the reactors. One job in a nuclear power plant is particularly hazardous. It's the job of the "sponges," unskilled short-term employees who expose themselves to relatively high doses of radiation for a brief time.

What Do "Sponges" Do?

They perform a nuclear plant's dirty work, thereby freeing skilled employees, whose annual exposure is limited by federal standards, to work longer hours. Sponges may be required, for instance, to squeeze through the eighteen-inch-wide passageways of a reactor's pressure vessel to simply tighten a bolt or weld a crack, where working briefly they may receive as much as 25,000 millirems of radiation an hour. Because the tunnels inside a reactor are narrow, and the shimmying through them often strenuous, sponges tend to be short, young men and women, whose ages make them extremely vulnerable to the genetic and somatic dangers of radiation.

Ten years ago sponges received 24 percent of the total dose at all nuclear power plants; today they are getting 50 percent of a higher overall dose. And more young people are being lured into the field by high salaries that vary with the risks of the job. NRC figures indicate that in the mid-to-late 1970s about 30,000 sponges were being employed each year.

The hiring of temporary employees to absorb radiation has been denounced for years by radiation health specialists. Even the NRC, which has yet to ban the hiring of sponges, admits that the practice flies in the face of a twenty-year-old federal radiation protection policy known as ALARA, for "as low as reasonably achievable" doses.

Then Why Does the Practice of Sponges Continue?

The NRC claims that the high radiation doses received by the industry's 50,000 full-time skilled employees has forced them to permit the hiring of sponges. Prior to TMI an NRC report did state that "efforts are being made to encourage the industry to abandon the practice of using extra workers and to rely instead on design and engineering efforts to reduce occupational doses." But within days of the TMI accident more than 100 sponges had volunteered to do whatever dirty work would be necessary at the plant.

While attempts are being made to correct reactor design failures that necessitate the hiring of such temporary workers, there is every indication that the use of sponges will continue — and increase — in the years ahead. We know that a relatively large dose of radiation received in one blast is potentially more dangerous than if the same dose is spread out over several weeks or months, permitting damaged cells to repair themselves. Sponges, who by the very nature of their job often receive 600 or 800 millirems a minute, may be potential time bombs for radiation diseases in themselves — and in terms of transmitting genetic defects to their children.

Don't Government Standards Limit the Amount of Radiation a Worker Can Safely Receive?

They do, for all people who work around radiation, whether they are power plant employees or hospital radiologists. But there is currently great confusion over just how safe the standards actually are. The standards issue concerns all of us, since radiation workers are of reproductive age and their defective genes can become part of the total human gene pool.

For most workers in nuclear technology, the Maximum Permissible Dose (MPD), excluding natural background exposure, is 5000 millirems a year. This is ten times what the MPD deemed safe for any individual in the general population, and more than thirty times the MPD for large communities of people. However, nuclear workers in certain specialized, essential jobs are permitted to receive 3000 millirems every four months for an annual MPD of 12,000 millirems. This is the highest safety exposure permitted under federal regulations, and many health specialists argue that it is dangerously high.

Certain organizations have already set their safety standards tighter than the federal ones. The Tennessee Valley Authority, the world's large nuclear proprietor with seventeen reactors in operation or under construction, has improved the safety standard for its employees by lowering the allowable radiation dose to 4000 millirems a year, and the Navy has adopted the same figure. Yet even these precautions may not be sufficient. In the last few years evidence has begun to accumulate indicating that workers exposed to low levels of radiation well beneath the federal safety standards have experienced increased incidences of leukemia and other cancers above those characteristic of the general population.

What Evidence Suggests the Safety Standards Are Unsafe?

In May 1980 the medical director of the Lawrence Livermore Laboratory, a weapons research facility in California, reported that twenty-seven employees there had contracted the rare skin cancer melanoma — an incidence three to five times higher than in the local population. The report is disturbing, but others are even more so.

In the May 1978 issue of the medical journal *The Lancet,* Boston hematologist Dr. Thomas Najarian published the results of his study of 592 deaths among workers in the Portsmouth Naval Shipyard, a naval facility in Kittery, Maine, where nuclear submarines are built, overhauled, and refueled. Najarian learned that nuclear workers suffered nearly double the percentage of cancer deaths and roughly five times the leukemia deaths in comparison with the general population and with non-nuclear workers at the same facility. The most startling statistics compared mortality among workers aged sixty to sixty-nine. In this group nearly 60 percent of the nuclear employees had died of cancer, while the cancer death rate among non-nuclear workers was only 26 percent.

Although the Navy had refused to release the radiation records on the shipyard employees, Najarian used published exposure figures for workers in related nuclear jobs and concluded that the average Portsmouth worker most likely received an annual dose of about 200 millirems — only twice the natural background exposure we all receive.

Has the Government Responded to Najarian's Findings?

Yes, a year after Najarian's study was published the federal government ordered the Navy to release the withheld radiation records to the Center for Disease Control in Atlanta, which is currently conducting its own study. The Department of Energy plans to study the six shipyards working on nuclear vessels and the residents who live near these locations.

Furthermore, fifteen years of radiation records on employees in the Navy's nuclear programs are currently being examined by Johns Hopkins University scientists in an attempt to determine the health hazards of levels of ionizing radiation not appreciably above the natural background dose. The Johns Hopkins findings, due in 1983, could strongly influence future safety standards for all occupationally exposed persons — especially if they confirm Najarian's findings concerning the danger of 200 millirem doses of radiation.

Isn't Two Hundred Millirems an Awfully Small Dose?

It is extremely small, and that's the reason for such concern. Najarian's findings indicate either that extremely small, continual doses of ionizing radiation are carcinogenic or that the radiation-measuring devices worn by workers are often inaccurate — or both.

The radiation devices worn by workers in the 1950s and 1960s mainly consisted of badges containing film that darkened when exposed to strong x rays or gamma rays. The badges, worn usually on a jacket lapel, were only a crude means of measuring radiation, and the further the irradiated body parts from the badge the less accurate the reading.

In addition, these older film badges, as well as various other detection devices, did not necessarily measure exposure to the alpha and beta particles emitted by most radioisotopes and nuclear wastes. In fact, shockingly little research has been conducted on internal radiation exposure as a result of inhaled alpha and beta-emitting particles, and whether this radiation correlates at all with badge readings. Najarian suspects that the subjects in his study probably received more radiation than their dosimeters indicated. Nonetheless, the threat to health from extremely small doses of ionizing radiation may be very real, as another recent study indicates.

What Was the Focus of the Study?

Four hundred and forty-two workers at the Hanford nuclear reprocessing plant in Washington. In the November 1977 issue of *Health Physics*, Dr. Thomas F. Mancuso, formerly under contract to the government on radiation issues, published results of his survey of the workers and found that those exposed to so-called "safe" amounts of radiation showed a 6 percent greater level of bone cancer, leukemia, and cancer of the pancreas than found in the general population. The average dose to those workers, all who died of cancer between 1944 and 1972, was 1000 millirems, merely one-fifth of the 5000 millirem Maximum Permissible Dose. Mancuso's figures indicate that the risk of radiation-induced malignancy is actually twenty times higher than what it is currently believed to be.

Mancuso had been hired by the government specifically to study the effects of low-level radiation on workers at the Hanford facility, but he was fired in 1976, and his duties were entrusted to the scientists at the Oak Ridge nuclear laboratory in Tennessee. Mancuso claims his alarming findings prompted his firing, and currently the Oak Ridge scientists have halted their study until an impartial panel of experts investigates Mancuso's claim.

Whatever the results of the investigation, one thing is clear: more research is needed into the health effects of radiation on workers. In fact, the Kemeny commissioners recommended just that in order "to provide a sounder basis for guidelines and regulations related to worker and public health and safety." The members also recommended that the states and utilities set up programs to educate workers and the rest of us about radiation-related health issues and procedures to help prevent any adverse effects of radiation on our health.

Has the Government Adopted These or Any Other Kemeny Commission Recommendations?

Of the forty-four recommendations, President Carter took action to implement thirty-eight, including many of those pertaining to worker safety, ensuring that the federal government will have a greater role in the way nuclear power plants are run. But Carter rejected the commission's major suggestion of abolishing the NRC and replacing it with an executive agency. While generally the report received widespread bipartisan support in Congress, there was little enthusiasm for such a change. Many congressmen, even those not considered friends of the nuclear industry, feared that by putting the agency in the executive branch the diversity of views provided by a multimember commission would be eliminated and the agency would be vulnerable to political pressure.

Recognizing the need for fresh leadership, President Carter did, however, replace the chairman who headed the agency at the time of TMI with another commissioner. And when a new agency opening occurred on the commission, Carter nominated a new full-time chairman from outside the agency. Carter also pushed for a five-member expert advisory committee to monitor the NRC's implementation of the Kemeny recommendations. In addition, he added a few of his own embellishments, including instructions to the NRC to accelerate its program of placing a resident federal inspector at every reactor site and a request for $50 million worth of increased funding for an expanded safety program.

Of course, even if most of the Kemeny recommendations are implemented and the new administration supports the Carter proposals, there is no guarantee that another Three Mile Island-type accident or worse won't occur. As the commission members themselves reported, "We have not found a magic formula that would guarantee that there will be no serious future nuclear accidents" — a point the foes of nuclear power are quick to seize upon in condemning nuclear power.

But Doesn't the Very Fact that Neither Three Mile Island Nor Any Other Power Plant Accident Has Resulted in a Meltdown Attest to the Safety of Nuclear Power?

Nuclear power proponents argue that it does, and claim that back-up safety systems are adequate to prevent a worst-case scenario. On the face of it their argument does sound rather convincing — if you ignore Three Mile Island that is. The expert consensus — governmental, scientific, and engineering — regarding TMI is that luck, Divine Providence, or both had more to do with preventing a major accident than any back-up safety systems. In fact, throughout the crisis such problems as leaks, stuck valves, and electrical mishaps underscored the inadequacy of the facility's back-up systems. In particular, one emergency back-up system designed to kick in immediately didn't because valve failure prevented the auxiliary feedwater line from opening.

Robert D. Pollard, a former reactor-licensing engineer for the NRC and now a staff member of the Union of Concerned Scientists, has raised a legitimate and crucial question concerning such safety equipment. "Can safety equipment," he asks, "operate in the environment created by the very accident for which that equipment is supposed to provide protection?" At Three Mile Island, safety equipment exposed to radiation inside the containment building failed.

What's more, one event at TMI — the formation of a mysterious hydrogen bubble — seriously calls into question the validity of fail-safe claims for reactor back-up systems, as well as risk-analysis studies concerning nuclear power.

Why?

Because formation of the bubble was never anticipated and therefore there were no back-up systems to eliminate the potentially dangerous build-up of pressure caused by the hydrogen. What's more, according to the director of the NRC's office of nuclear reactor regulation the presence of the gas bubble was "a new twist" which had never been considered in any of the governments' computerized accident simulations.

For days following the accident officials feared that the bubble might expand, causing the fuel to reheat, which could result in a meltdown. Although subsequently the NRC said it had overestimated the danger of the bubble, the fear of a meltdown forced nearly 60,000 area residents to flee their homes. It took two days to analyze the situation and another three to get it under control. Finally, through a complicated process the gas was released into the atmosphere and an accident was avoided.

But the Chance of a Serious Accident Exists with Almost All Energy Sources, and for That Matter with Much of Modern Technology?

Yes it does, and this is a point the utility companies and nuclear industry people stress as well. For instance, it is true that thousands of people die each year as the result of pollutants from the combustion of coal. In fact, according to a congressional report the number of coal-related deaths is about 48,000 a year, a high percentage of which are miners. In addition, another argument of the nuclear industry is that thousands of people die on our highways every year, and that other industry-related accidents take the lives of countless more. No civilians, they point out, have yet to be killed by a nuclear power plant. There certainly is no denying these facts, and on the face of it the arguments seem persuasive enough. But the logic that underlies them grossly distorts the true picture.

Why Is This Logic Faulty?

For several reasons. First, there is the simple fact that despite the number of deaths coal-generated electricity, automobiles, or any technology may be responsible for, the potential for death and destruction from nuclear power generation is fantastically greater. Even in the worst conceivable accident at a coal-powered plant the results are localized health consequences, and possible death and destruction on a small scale, not tens or hundreds of thousands killed with the possibility of genetic consequences for generations, plus large tracts of land laid waste for years.

What's more, a comparison of nuclear power and coal-related death statistics may well be misleading. Every nuclear power plant emits some radiation whether it's from leaking pipes, vented gases, or waste water. While these emissions are small and considered harmless, no one has tallied how many illnesses or deaths they may have been responsible for. The health effects of such radioactive emissions, particularly the kind that occurred at TMI, are unlikely to be resolved — at least not for decades.

In addition, the largest percent of coal-related deaths occurs among miners and that is their profession by choice. We also choose whether or not we wish to risk driving or riding in an automobile. Who of us has had such a choice when it comes to the location of a nuclear plant near our homes?

Still, Why Fear the Hypothetical, When So Many Experts Feel the Odds Against a Worst-Case Nuclear Disaster Remain Very Slim?

Since Three Mile Island, the improbable seems less so. But for anyone who thinks the odds are really that slim, it may be sobering to know what our government is doing just in case the "impossible" happens. In 1977 the Federal Preparedness Agency was given the task of drafting an interim guidance plan for dealing with a new version of the unthinkable — a peacetime nuclear emergency, whether it be touched off by an accidental bomb explosion, or sabotage, or a meltdown at a nuclear power plant.

In the event of a major accident, the plan calls on the Department of Health and Human Services, the FBI, and the Postal Service to identify the dead and attempt to reunite families. The health department is also to establish mental-health centers and condemn or embargo food if necessary. The Agricultural Department is to dispose of contaminated livestock and issue food stamps if warranted, and the Environmental Protection Agency is to devise plans for decontaminating water, food, and buildings, and decide how to dispose of the dead, who could be highly radioactive. And in the event that a nuclear accident threatens a large segment of the population, the Justice Department is responsible for restoring any civil liberties that might be suspended because of a breakdown in the social order.

In 1979 the government went a step further and handed over to a special body, the Federal Emergency Management Agency (FEMA), the responsibility for dealing with peacetime nuclear emergencies. FEMA is drafting its own set of emergency guidelines and taking the Federal Preparedness Agency's interim guidance plan into consideration in the process.

Granted It's Frightening, but the Government Can Hardly Be Faulted for Being Prepared.

That's not the point. The real issue is why the government is pursuing a power source fraught with such potential danger. Even if such emergency requirements had to be instituted on the most limited scale, in an area surrounding one power plant, the economic consequences could be far-reaching. Since Three Mile Island — which was only a near–nuclear disaster — serious questions have been raised about the viability of nuclear power, and progress toward greater reliance on nuclear energy has been slowed, with a virtual de facto moratorium on power-plant construction. Imagine the consequences for the nuclear industry if a worst-case accident occurred. Surely an outraged citizenry would demand a shutdown of many, if not all, plants. What kind of an energy bind would we then be in, and what kind of wrenching adjustments would we be forced to make in our standard of living, especially if we have dramatically increased our dependence on nuclear power? In this respect, nuclear power is only as strong as its weakest link.

Yet even barring a major accident, nuclear power is a risky source of power for another reason — its cost makes it a highly unreliable source of energy for the future.

But Isn't One of the Major Assets of Nuclear Power the Fact that It's a Cheap Source of Energy?

Nuclear-power proponents have traditionally argued just that. From the earliest days of nuclear power, industry and government officials claimed that the economics of nuclear energy made it by far the cheapest source of electricity — so inexpensive, in fact, that one day it would be "too cheap to meter." Perhaps that was the way it appeared in the 1950s and 1960s, but since then nuclear-power-generated electricity has become increasingly expensive and less competitive with other sources of electricity — especially coal.

Though nuclear-generated electricity is cheaper than the electricity we derive from oil, it is now at least as expensive as coal-generated electricity. According to DOE statistics, for each of the past three years the cost of generating electricity from nuclear power has jumped about 15 percent, while the cost of coal-generated electricity has remained fairly stable. As a result, the 35 percent cost advantage that nuclear-generated electricity had over coal in 1977 dropped to 30 percent in 1978 and was entirely wiped out in 1979. When the figures are in for 1980, the cost of coal-generated electricity is expected to at least hold its own — if not actually turn out to be cheaper than electricity derived from nuclear power, as some forecasts indicate.

What Basically Accounts for the Rising Cost of Nuclear Power?

Increased generating expenses are partially responsible, but primarily the price of nuclear power is on the rise because of spiraling construction costs due to multiplying safety regulations and litigation brought against the utilities by environmentalists and citizens' groups.

Early nuclear power plants required a minimum of safety systems. But as plants have grown larger and necessitated more sophisticated safety systems, and environmentalists have demanded much-needed additional safeguards, power plant costs have soared. The safety systems themselves are costly, but on top of this the NRC is often forced to delay the licensing of a plant for years because of court action taken by groups opposed to plant construction. The result has been huge cost overruns fueled by rising inflation and interest payments on the nearly completed but unlicensed plants. A delay of one year adds an estimated twenty to thirty million dollars to a plant's price.

A comparison of plant construction costs over the past decade underscores the rising price tag of nuclear power. In 1967 the cost of bringing a typical 1000 megawatt nuclear generating unit into operation was about $34 million. At the time of the TMI accident the cost had jumped to $1 billion. Utility and industry officials believe that as a result of TMI utilities will have to spend tens to hundreds of millions of dollars more on a nuclear plant because of more frequent shutdowns, design and safety changes, and licensing delays. Their real fear is that construction delays will put plant costs entirely out of reach. Already one plant in Michigan begun fifteen years ago is costing more than $3 billion to build, and it won't be completed for several more years. Numerous other plants have passed the several-billion mark as well.

In addition to increased construction costs there is yet another reason why nuclear-generated electricity has proved less economical than it was once thought to be — the "capacity factor" for nuclear power plants has been less than anticipated.

What's the Capacity Factor?

Simply the percentage of full capacity at which a plant operates. Initially, the AEC, along with reactor manufacturers, predicted that nuclear power plants would produce electricity over their thirty- to forty-year life span at 80 percent capacity. But that optimism was not realistic. The NRC began compiling capacity factor data on nuclear plants in 1973. In that year the nation's large nuclear power plants recorded an average capacity factor of 38.4

percent. By 1978 the figure had climbed to 67 percent, but it dropped to 60 percent in 1979. While this is comparable to coal-fired facilities of the same size, it is considerably less than industry expectations. Utility officials claim that the most recent statistics are misleading, because the accident at TMI forced the temporary shutdown of several reactors, and they point out that the newer reactors coming on line will have increasingly fewer problems as technology is perfected. Such claims, though, are themselves misleading, for shutdowns will surely continue to occur, and they must be factored into any projections of the future capacity and cost competitiveness of nuclear energy.

There is no denying that generating and capital costs are on the rise for all energy sources, but capital costs are running particularly high for nuclear power, and currently exceed those for coal. By some estimates the average difference between nuclear and coal capital costs increased from 5 percent to 50 percent during the 1970s. If past relationships continue, nuclear power plants could cost at least 50 percent more to build than coal plants by the late 1980s, and if that happens, nuclear generating costs will substantially exceed those of coal. This would surely have an impact on the pace of power-plant construction, which has already been slowed largely as a result of the "uneconomics" of nuclear power.

Has There Been Much of a Slowdown?

A dramatic one. In 1974, for instance, the government had planned on building between 600 to 720 reactors by the year 2000. Today, that figure has been pared back, by even the most optimistic estimates, to about 400, with the government forecasting only about 150 to 200 reactors by the turn of the century. Granted, slowed economic growth and less rapidly rising electricity demands account for some of the reduction, but the costs of nuclear power are the major cause.

The slowdown has been worldwide. Even though several heavy energy importing countries like France are vigorously pushing nuclear power, in the last six years only one-third the number of reactors originally planned worldwide have been built.

In this country, orders have so slackened that reactor manufacturers are living largely off their still considerable reactor backlog, as well as their profitable fuel fabrication and maintenance contracts. Conventional wisdom has it that a supplier needs between four and six orders each year to remain economically healthy. However, in recent years cancellations have far outpaced orders, leaving each of the companies' reactor manufacturers with fewer than four orders each year. Company officials acknowledge that if the current trends aren't reversed the nuclear industry could be wiped out. The major reactor manufacturers, Westinghouse and General Electric, and smaller suppliers like Babcock and Wilcox, the company that produced the two reactors at TMI, feel that they can tolerate several more years of slow business if orders pick up again by the mid 1980s and if plant lead times are cut back from ten to fourteen years to seven or eight years.

That Sounds Unrealistic?

Given current construction delays and costs it is, and it's hard to imagine how manufacturers or industry officials could possibly believe that reactor orders will substantially increase at least in the near future. If anything, in the wake of TMI the public has been sensitized to the nuclear power issue and more, not less, litigation is likely. Plus more and stiffer safety regulations are sure to lead to further delays, and ensure that the slowdown in reactor construction continues.

The slowdown, though, has been a plus for the nuclear industry in one respect; it has kept the price of uranium fairly stable. Recently there were widespread fears that increased fuel demands would push the price of uranium from around $40 per pound to about the $100 mark, and there was talk of the formation of a uranium cartel similar to oil's OPEC cartel that might trap those countries lacking sufficient uranium supplies in an energy noose. Now if the present building trend continues, demand in the year 2000 may be no more than a third of what seemed likely only a few years ago, and a uranium shortage would be the least of the industry's worries — at least for the short term.

*If Nuclear Power Is Pushed More Aggressively, Those Earlier
Fears of a Uranium Shortage May Soon Become a Reality?*

That's true. Whether you are more pessimistic about the rate of
power plant construction or more optimistic about uranium resources, sooner or later problems with uranium supplies are likely
to occur. Most experts agree that given the current rate of power
plant construction, there will be a serious shortage of uranium
before the year 2020; obviously a quickened pace of construction
could result in a shortage even sooner.

If all the reactors in this country remain the current "light
water" kind, it is estimated that the country's total uranium ore
resources would be sufficient to fuel only 600 reactors for forty
years. In addition, all of the known low-cost uranium reserves
have already been contracted for. And if uranium reserves are as
low as experts suggest, we would eventually be forced to depend
on foreign supplies and could possibly face a uranium cartel. Industry officials, however, argue that we won't have to worry
about a uranium shortage if we rely on breeder reactors.

What's a Breeder Reactor?

Basically, a breeder reactor differs from a conventional "light
water" reactor in that it uses chemicals such as sodium instead of
water as a coolant and it can use plutonium as well as uranium for
fuel. The breeder can extract energy not only from the rare isotope
uranium-235, which is used for fuel in light-water reactors, but
also from low-grade ore of other isotopes of uranium. In doing so,
it can increase the supply of uranium sixfold. The breeder would
accomplish this mainly by using uranium-238, which makes up
99.3 percent of natural uranium, and which can be converted into
plutonium; in conventional reactors only 1 percent of the uranium
mined can be used. It is estimated that with breeder technology
there would be enough uranium in this country to supply 1000 reactors for 40,000 years.

According to breeder proponents there is another advantage to
the reactor: because light-water reactors generate plutonium as a
waste product, they offer us a good source of fuel for breeders. So

the waste from the one plant could become the power source for the other. In addition, the breeder reactor produces more fuel than it consumes — hence the term breeder — which can be recycled into fuel for light-water reactors. Breeder proponents argue that this potential symbiotic relationship offers hope of an almost inexhaustible source of energy.

Are There Any Breeders in Operation?

In various countries there have been numerous experimental models in operation for some time, beginning with the first ever to produce electricity, the EBR-I, which began operation in Idaho in 1951. The most recent, as well as the most advanced and ambitious, experimental breeder reactor to go into action in this country is the Fast Flux Test Facility at the Hanford plant in February 1980.

Despite advances in breeder technology, numerous questions remain to be answered about its commercial feasibility, and preliminary experiments have proved disappointing in terms of the reactor's doubling time — the amount of time it takes to double the original inventory of fuel — and the breeder may prove not to be an energy panacea after all.

Even if all goes well, commercial breeders are not expected to become a reality until the mid 1990s. But if breeders do prove feasible, antinuclear groups argue they should not be part of our energy future.

For the Same Reason They Object to Conventional Nuclear Reactors?

Yes, but even more importantly because they charge that breeders would increase the world supply of bomb-grade plutonium and, therefore, encourage the spread of nuclear weapons and increase the likelihood of plutonium falling into terrorist hands.

In 1977, President Carter proposed a conference of technical experts from industrialized and developing nations to examine the proliferation problem. The recently released study concluded that breeders pose no greater risk of nuclear-weapons spread than existing power plants now in use in the United States and many other countries. Whether or not this is the case — and many experts say it is not — the report does paint an ominous picture regarding plutonium growth, based on projections for the future growth of nuclear power. If these projections prove correct, the world's supply of bomb-grade plutonium would increase sevenfold to a total of about 500 tons during the 1990s. A frightening amount when you consider that it only takes about twenty-five pounds of so-called commercial-grade plutonium to produce a single nuclear weapon.

The potential consequences of such a vast lethal plutonium supply has prompted the International Atomic Energy Agency to push for the establishment of internationally controlled plutonium storage banks where countries would deposit their "excess" stocks of weapons-grade material, which they could retrieve when necessary. The idea has the backing of many countries including our own, and a storage agreement is expected to be in place soon, allowing IAEA inspectors for the first time to directly control weapons-grade material.

A Country Might Be Able to Fashion a Bomb from Plutonium Waste, but Wouldn't It Be Extremely Difficult for an Individual to Make One?

It is impossible for an individual, or even a few individuals, to make a *military type* bomb, whether it be from weapons-grade or nonweapons-grade plutonium. The difficulty is compounded with nonweapons-grade material because of the complex and expensive engineering, chemical, and metallurgical techniques required to convert the plutonium and uranium to an enriched weapons-grade metal.

But crude nuclear bombs can be designed by a person with a knowledge of physics equivalent to that of a second-year college physics student. And the materials needed to construct the bomb can be rented and obtained in a well-stocked hardware store for a few thousand dollars. More frightening, though, is the fear that a much simpler nuclear device could possibly be made which would require no calculations and no special tools. Some experts believe that such "kitchen bombs" can be made from a bucket of pluto-nium oxide, a powder extracted from power plant waste, that could be detonated with ordinary high explosives. The crude bomb would not devastate a city, but it could, for example, knock the top hundred floors off the Empire State Building, and the ra-dioactive fallout could leave deadly debris — particularly pluto-nium in the lungs of many people, even suburbanites, in the vicin-ity of the blast.

Despite the Dangers Inherent in Nuclear Power, Isn't It an Essential Source of Power?

Currently this country's nearly seventy reactors supply about 12 to 13 percent of our electricity needs, depending on how many are operating at any given time. Nuclear industry officials as well as other nuclear power proponents, both in and out of government, would have us believe that not only are we absolutely dependent on these seventy reactors for our existing energy needs but that there is no way we can meet our future energy requirements with-out at least tripling or quadrupling the existing number of reactors in the next twenty years. Their reasoning is based largely on the familiar arguments: that oil supplies are rapidly diminishing; that oil prices are skyrocketing; that increased reliance on coal is unrealistic because of environmental objections; that energy use in the coming decade will continue to soar; and that alternate energy technologies, like solar, are in too primitive a stage to supply our energy needs for the foreseeable future.

Those arguments do sound convincing, but except for the esca-lating price of oil they are largely wrong. For instance, it is true that the demand for electricity continues to grow in this country, but it is growing less rapidly than in the past. Since 1974, Ameri-cans have been using far less electricity than utilities had expected. Until the early 1970s electricity consumption rose about 7 percent a year. But in 1979, it dramatically dropped to 2.8 percent. Cur-rently it is growing at about 1.4 percent and is expected to decline even further in the foreseeable future.

While recent milder winters account to some degree for the lower consumption figures, according to many experts the numbers represent a fundamental change in our appetite for electricity; specifically, soaring oil costs have apparently made many of us more electricity-cost conscious and willing to conserve.

The dramatic impact such conservation has had on electricity consumption contradicts what pronuclear forces tell us; that even with a major concerted national conservation effort we will need many more nuclear power plants by the turn of the century. By the most optimistic projections, if current electricity consumption rates continue, the nation's utilities would not have to order any more nuclear power plants to meet future growth needs. So you can imagine the impact a government-legislated conservation effort would have on nuclear power.

Yet barring that, and acknowledging that current growth projections could be wrong, this country has sufficient coal to meet its energy needs. We have nearly 435 billion tons of coal capable of yielding more energy than all the oil in the Middle East — enough to last for hundreds of years. Energy officials in and out of government do not deny that our vast coal resources could be mined to satisfy most of our growth in electricity demand through the end of this century.

Isn't Coal Too Dirty to Rely On as an Energy Source?

Not necessarily. With the proper environmental safeguards coal can be a relatively clean and reasonably cheap source of energy. It is already more competitive than oil and is becoming increasingly competitive with nuclear power.

However, an important point must be made: even if coal-generated electricity were not competitive with nuclear-generated electricity, that should be no reason for us as a nation to choose nuclear power over coal power. It is safety, not cost, that should primarily determine which energy route we pursue.

The increasing competitiveness of coal has recently been underscored by an internationally sponsored World Coal Study compiled by Professor Carroll Wilson of the Massachusetts Institute of Technology and experts from sixteen countries that produce and use most of the world's coal. The study concluded that oil prices are so high that it is now possible to spend heavily to clean up coal and still come out ahead in meeting our energy needs. The report contends that without sacrificing health, safety, and environmental protection coal can supply the world's energy needs for at least two decades, sustaining moderate world economic growth.

What About the "Greenhouse Effect" from Increased Coal Burning?

The possibility that the carbon dioxide produced by burning coal — as well as other fossil fuels — could have a "greenhouse," or heating, effect on the earth causing catastrophic changes in global climate cannot be ruled out. Whether or not such climate changes, though, would surely result has been hotly debated by scientists. Even among those who argue the climatic threat is real, many question whether the burning of coal, in addition to the other fossil fuels we already use, would really make that much difference. What's more, it would take many decades, perhaps centuries, for the cumulative effects of coal combustion to cause disruptive climatic changes. Long before then, new and improved technologies would enable us to cut back on our reliance on coal.

But the whole greenhouse-effect argument against coal may be moot. It is not necessary for us to rely heavily on coal indefinitely, or even until it might cause us harm. We need only use it intensively for the short-term, to buy ourselves time — about twenty years or so — the time that is necessary to allow for the development of safe energy alternatives. Synthetic fuels and biomass — organic fuel from sources like sugar cane and other vegetation — are already showing great promise, and in time they along with solar power and possibly fusion will adequately supply our energy needs.

What's Fusion Power?

Fusion is the energy process that fires our sun, and it's nature's way of generating power. Fusion has two major advantages over fission, the process that fires our present nuclear power plants: fusion produces amazingly little radioactive wastes — in fact it's estimated to be 1000 times cleaner than fission — and fusion is fueled by hydrogen, the most abundant element in the sea, so there would never be a danger of a fuel shortage.

Technically, it's a lot harder to harness fusion and that's why fission became the preferred choice in nuclear power plants. Where fission involves the splitting in half of large atoms of uranium, fusion depends on the joining together of small atoms of hydrogen. To achieve fusion a hot gas, or plasma, of hydrogen atoms must be confined securely into a small space, squeezed to enormous density, and heated to 100 million degrees Celsius. Despite such difficult requirements, several groups of scientists are making promising advances in achieving a fusion reaction in the laboratory, and they project that fusion power could be a commercial reality in twenty years. Already a group of engineers from government laboratories and private industry are engaged in a study, called Starfire, aimed at the design of what may become the prototype for our first commercial fusion reactor. Researchers in the field agree that in theory no single energy source on earth has more potential than fusion for powering our world in the twenty-first century and beyond.

With Increased Reliance on Coal and Newer Energy Forms, Could We Safely Abandon Nuclear Power?

We certainly can, and it's the only sane, sensible thing to do. But we must do so gradually. If we were to shut down all nuclear reactors in this country overnight, there is no way we could realistically compensate for the energy shortfall — at least not without seriously disrupting the economy, since no alternative energy plan has been developed. What the government should do is declare a permanent moratorium on all nuclear power plant construction, permitting only those reactors currently operable to produce electricity. And as soon as the development of other energy sources allows, the existing reactors should be phased out. The goal should be to abandon nuclear power no later than twenty-five to thirty years from now — when many plants will already have reached their life expectancies.

To accomplish this the government will have to commit itself on the scale of the Apollo moon program to meet our immediate energy needs with coal and our future energy needs with other promising forms of energy such as solar and fusion power.

In the meantime, to ensure maximum safety Washington must adopt and strictly enforce the Kemeny commission recommendations that apply to existing power plants, in particular, the periodic renewal of plant licenses, the placing of federal inspectors at each facility, the establishment of government-certified training centers for plant operators, and the upgrading of safety systems.

Beyond these actions, the government should also transfer plant security to U.S. military personnel or a federal guard service. And short of abolishing the NRC, there should be a thorough house cleaning of the agency to rid it of the problems that helped contribute to Three Mile Island. In addition, Congress must move quickly to amend the Price-Anderson act, which since 1975 has strictly limited the public liability of nuclear power plant owners in case of an accident. Under the act, total assured protection per accident to the public is only $560 million, to be shared by the plant owners and the government. Such coverage is grossly inadequate; damage estimates for even a precautionary evacuation of the Harrisburg area at the time of TMI, ranged from a low of $3 billion to a high of $17 billion.

Only by adopting such measures can we be assured of relative safety and protection while we move toward more promising, much safer energy sources to meet our needs and those of future generations.

Appendices

APPENDIX A

Average Doses for Diagnostic X ray Examinations in Millirems*

Test	Number of films	Dose to local skin	Dose to bone marrow	Dose to gonads Men	Women	Estimated annual excess cancer and/or leukemia deaths per million people exposed
Skull						
Cervical spine (neck)	4	1200	80	1	4	3–9
Dental whole mouth	4	600	50	<10	<10†	2–8
Shoulder	4	4400	20	<1	<1	1
Chest (film)	2	400	60	1	1	3–7
Chest	2	90	15	5	8	1
	Fluoroscope	1000	200	<10	<10	4–10
Thoracic (middle) spine	3	3600	250	30	30	7–9
Lumbar (lower) spine	3	6000	360	2268	275	7–9
Abdomen	3	2100	120	254	289	5–25
Mammography	3	4500		—	1	5–20
Upper gastrointestinal series	Fluoroscope	1550	1600	137	558	40–120
Lower gastrointestinal series, or barium enema	Fluoroscope	1600	1800	1585	805	30–100
Cholecystography (gall bladder)	3	2100	350	10	200	20–70
Intravenous pyelogram or IVP (kidney, ureter)		1200	400	300	500	7–50
Pelvic	3	1450	420	300	300	6–22
Hip or upper femur (thigh bone)	3	3000	100	1064	309	6–22
Arms and lower legs	2	200	20	<10	<1	1

* The doses in the table are averages. Actual exposures from diagnostic tests can vary considerably, depending on the training of the technician or doctor taking the radiograms, the safety precautions followed to protect the gonads and blood-producing organs, and the accuracy of the x-ray machine. Also, the estimated annual excess cancer and leukemia deaths per one million people exposed are for adults; the figures would be considerably higher for children and infants who received these tests.

† The sign < means "less than."

APPENDIX B

Cancer Risks from X rays

Cancer	Number of expected cases for one million people exposed to 1000 millirems
Breast	50 to 200
Thyroid (external exposure only)	50 to 150
Lung	50; it can be significantly higher for smokers
Leukemia	15 to 25
Bone tumors	2 to 5
All other organs	Less than one

APPENDIX C

Radiation Doses to a Fetus from Diagnostic X ray Tests

Test	Dose in millirems *
Abdominal Series	100 for each x-ray film; up to 2000 with fluoroscopy.
Aortography (aorta)	6000 to 20,000 depending on technique employed.
Barium Enema (large intestines)	350 to 6000; higher with fluoroscopy.
Cervix	Up to 6000 if fluoroscopy is used.
Heart Series	5 to 50; higher with fluoroscopy.
Celiac Angiography (abdominal blood vessels)	2000 with film; up to 20,000 with fluoroscopy.
Chest	Less than 10 with film; 5 to 70 with fluoroscopy.
Cholangiography (bile ducts)	20 to 200 with films; up to 2,000 with fluoroscopy
Cholecystography (gall bladder)	20 to 200 with films; up to 2000 with fluoroscopy
Cystogram (bladder and urethra)	500 to 1000
Dacryocystography (tear ducts)	1 to 10
Dental Series	0 wearing lead-rubber apron.
Esophagram (esophagus)	Up to 500 with fluoroscopy.
Extremities	1
Lower GI Series	350 to 6000; higher with fluoroscopy.
Upper GI Series	100 to 2000; higher with fluoroscopy.
Hip, Spine, & Buttocks	300 to 700

Test	Dose in millirems*
Hysterosalpingography (uterus, Fallopian tubes)	1200 to 6000, but rarely performed on a pregnant woman.
Ileo-Cecal Study (ileum)	100 to 2000; higher with fluoroscopy.
KUB (kidney, ureter, & bladder)	200 to 6000
Myelography (spine)	2000
Nephrotomography (kidney)	500 to 2000
Obstetric Exam	500 to 2000
Ocular (eye)	1 to 10
Pelvic Pneumography (Pelvis-lungs)	300
Pelvimetry	600 to 4000
Placenta Praevia	500 to 1000
Placentography	300 to 7500
Pneumoencephalography (brain)	2 to 20
Polytomography (both ears)	3 to 30
Pyelography, Intravenous (IVP; kidney, ureter)	400 to 2000
Renal Arteriography (arteries near kidneys)	2000 to 4200
Salpingography (Fallopian tubes)	10 to 50
Shoulder	Less than 1
Sinus Series	1 to 10
Spine, Lumbar (sacrum & coccyx)	200 to 2000
Spine, Dorsal (cervical & thoracic)	1 to 3000
Splenoportography (spleen)	2000

* Doses vary with the number of films taken, the use of fluoroscopy, and other radiographic procedures. A doctor will always try to keep the dose as low as possible since the fetus is highly vulnerable to radiation damage.

APPENDIX D

The Effects of Radiation on the Human Fetus*

Deformity	Gestation days fetus is most vulnerable
Anencephaly (no brain)	20 to 30
Exencephaly (exposed brain)	15 to 25
Microcephaly (small brain)	15 to 25
Hydrocephaly (swollen brain)	25 to 35
Pituitary gland	32 to 41
Lower jaw	12 to 18
Cleft palate	28 to 34
Teeth	30 to 38
Ear	20 to 30
Nose	20 to 30
Spinal curvature	25 to 35
Eye	20 to 30
Cataracts	1 to 6
Abdominal hernias	12 to 20
Liver	20 to 70
Fingers and toes	20 to 45
Skeleton	9 onward
Mental retardation	16th to 20th weeks
Stunted growth	16th to 20th weeks

* These estimates, from data by Dr. Roberts Rugh, are based on research with mice. In terms of fetal development, there is reason to believe that indications of deformities in mice fetuses can be extrapolated with more than a fair degree of accuracy to human fetuses. The figures in the table represent periods of maximum vulnerability to radiation, but many deformities can occur beyond the cutoff times indicated. Notice that most congenital abnormalities are induced between the 20th and 40th days of gestation.

APPENDIX E

Radiation Dose from Diagnostic Tests That Use Radioisotopes

Test	Radioisotope	Average dose in millirems to target organ	Average dose in millirems to whole body
Lung Scan	Iodine-131	2000	120
	Technetium-99m	750	10
	Indium-113m	1000	24
	Xenon-133	375	1
Thyroid Scan	Iodine-131	75,000	200
	Iodine-125	65,000	60
	Iodine-123	1500	3
Liver Scan	Gold-198	6000	170
	Technetium-99m	400	14
	Indium-113m	750	22
	Iodine-131	800	300
Pancreas Scan	Selenium-75	3500	1750
		7000 (liver)	
		2000 (gonads)	
Spleen Scan	Chromium-51	7000	60
Bone Scan	Strontium-85	3800	115
	Strontium-87m	300	40
	Fluorine-18	260	50
	Technetium-99m	450	100
Kidney Scan	Mercury-197	1500	15
	Technetium-99m	50–7000 *	8–120
	Iodine-131	300	90

* The actual dose depends on the chemical substance the radioisotope is mixed with.

APPENDIX F

Radiation Dose to Mother and Fetus from Diagnostic Tests That Use Radioisotopes

Test	Radioisotope *	Whole body dose in millirems to mother	Whole body dose in millirems to fetus	Dose in millirems to fetus's thyroid gland †
Thyroid Scan	Iodine-131	27	15	5000
Plasma Volume	Iodine-131	17	10	5000
Liver Scan	Iodine-131	36	19	5000
	Technetium-99m	30	—	—
Brain Scan	Technetium-99m	100	—	—
Placentography	Iodine-131	15	7	4900
	Technetium-99m	9	2	10

* Whenever possible, iodine-131 should not be administered to a pregnant woman. Beginning in the tenth week of gestation a fetus's thyroid gland accumulates iodine, and at a faster rate than the mother's gland. It requires a very small dose of radiation to induce thyroid cancer in a fetus, and the disease usually appears years later. When a radioisotope test on a pregnant woman is imperative for her health, an alternate nuclide, technetium-99m, should be used. It is much less hazardous because of its short half-life of six hours — compared to iodine's half-life of eight days — and it emits only gamma rays, whereas iodine emits both gamma rays and beta particles.

† Radioisotopes are often combined with drugs or proteins to facilitate their journey through the body to the organ being studied. Some proteins and drugs do not reach the fetus because their molecules are too large to pass through the membrane of the placenta; others do. This accounts in part for differences in whole body doses to fetuses from the same radioisotope.